D0435439

Everyday Math Made Easy

Peter Davidson

McGraw-Hill Book Company

New York St. Louis San Francisco Auckland Bogotá
Guatemala Hamburg Johannesburg Lisbon London
Madrid Mexico Montreal New Delhi Panama Paris
San Juan São Paulo Singapore Sydney Tokyo Toronto

First McGraw-Hill Paperback Edition, 1984

1 2 3 4 5 6 7 8 9 SEMSEM 8 7 6 5 4

ISBN 0-07-049628-5

LIBRARY OF CONGRESS CATALOGING IN PUBLICATION DATA

Davidson, Peter,
Everyday math made easy.

1. Mathematics—Handbooks, manuals, etc.
I. Title.
QA40.D38 1984 513′.93 84-9692
ISBN 0-07-049628-5

Book design by M. R. P. Design.

Dedicated to two very special people, Kim and Neil, in honor of their very special occasion.

Peter Davidson

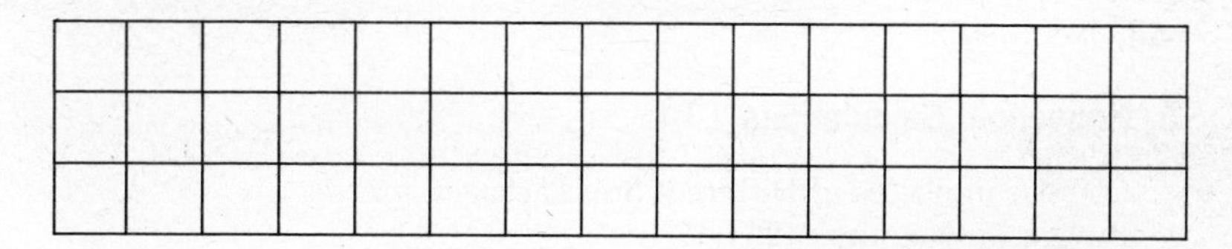

Contents

3. Household Calculations / 21

4. Energy and Utilities Calculations / 31

5. Building and Decorating Calculations / 63

Everyday Math Made Easy

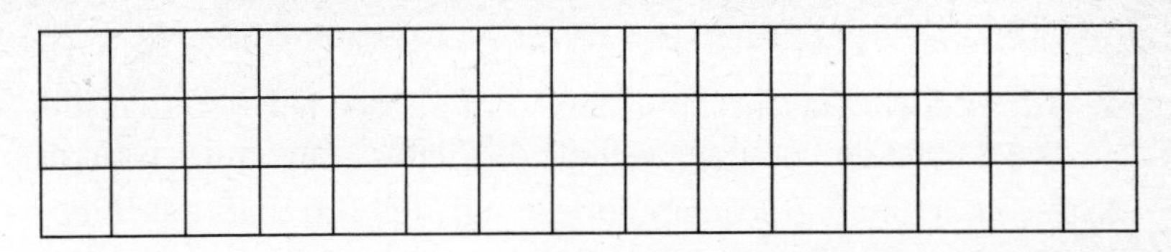

Introduction

Try your hand at answering these questions:

1. You are considering the purchase of a home, for which you will need to borrow $50,000.

a. What will be your monthly payments if you obtain a 25-year loan for $50,000 at 12% interest?

b. What will be your total payments over the 25 years to pay off that $50,000 loan?

2. Jannsen Carpets, Inc. is offering carpet at a sale price of "only" $7.50 per yard. The carpet in your living room, which measures 12 feet by 18 feet, just happens to be both out of style and threadbare.

a. How many square yards of carpet will it take to carpet your living room?

b. How much will the carpet cost if you buy it at Jannsen's sale price?

3. You make an error in calculating your checkbook balance and, consequently, your account becomes overdrawn by $2. Your bank honors (pays) the check which caused the overdraft and charges you a $10 overdraft fee for doing so. On the day following the overdraft, you make a deposit to your account. What rate of interest does the $10 fee charged by your bank represent?

4. A friend who has just returned from Europe tells you that she drove 600 kilometers in a single day. How many miles is that?
5. A couple of your acquaintance smugly tell you that their Electroenergy stock went up $1\frac{1}{4}$ points per share yesterday. How many dollars increase is that?
6. A vacant lot measuring 110 feet by 132 feet is for sale for $20,000.
   a. How many acres is the parcel?
   b. What is the price per acre?
7. An article in your local newspaper states that your real estate tax will be 2 *mills* higher next year. How much additional real estate tax will you have to pay if your home has a $60,000 assessed valuation?
8. Assume your local utility company charges 6¢ per kilowatt-hour of electricity used. How much does it cost your family to needlessly burn a 100-watt light bulb an hour a day, each day, for a 365-day year?
9. You apply for a personal loan at your local bank, and the loan officer tells you that you must submit a personal *balance sheet* before your loan application can be processed. Do you know what possessions to include, how to place a value on your possessions, and how to complete the balance sheet?
10. You are thinking about adding a second income either by moonlighting at a part-time job yourself or by having your spouse take a job. How much real, spendable income will that second job add after the additional income taxes, Social Security tax, and other expenses related to that job are paid?

These questions and similar ones seem to pop up frequently in our day-to-day lives. When you encounter one of them, can you quickly and accurately perform the calculations to answer it? If you are like most people, your answer is "I'm not sure," "You've got to be kidding," or "Heavens, no!"

Everyday Math Made Easy was developed to serve as a guide for performing the calculations that come up unpredictably throughout life. A short but thorough description is presented for each type of calculation, and then an example and solution are provided to lead you step by step through the process of performing the calculation. It's so easy that you can get the right answer the first time—even if you think your present mathematical abilities are rather shaky!

Everyday Math Made Easy is a handy guide that will prove to be most helpful for years to come. Using it will help you better understand your circumstances. It will provide the assurance and peace of mind that come from *knowing* that you have made an accurate calculation and *knowing* where you stand. Keep *Everyday Math Made Easy* within handy reach. You will find that it will help make your everyday living easier!

By the way, the answers to the first eight questions listed above are as follows:

1. ***a.*** $526.62; ***b.*** $157,986.00

2. ***a.*** 24 square yards; ***b.*** $180

3. 182,500% (Yes! 182,500%)

4. 360 miles

5. $1.25 per share

6. ***a.*** One-third acre; ***b.*** $60,000 per acre

7. $120

8. $2.19

Well, how many did you get right? If you missed none or only one, you get an A. If you missed more than one, there is an easy solution: *Everyday Math Made Easy*!

Peter Davidson

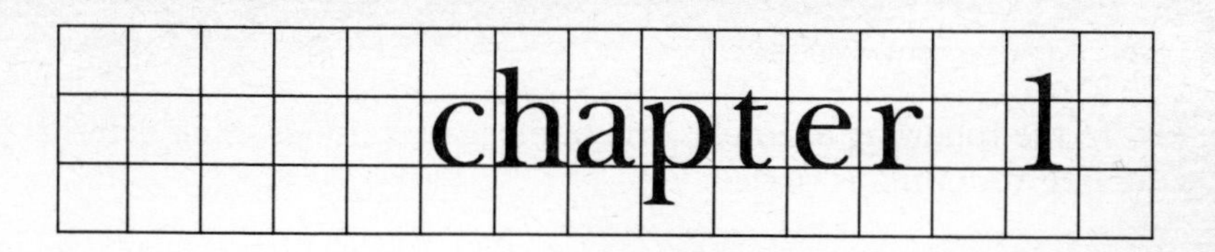

A Very Brief Math Review (We Promise)

This brief review will help you over any mathematical stumbling blocks you might encounter in performing the calculations in this book. Also, some of the techniques will help you perform your computations much more quickly and accurately.

Rounding Numbers

Numbers are rounded by converting them to the nearest cent, dollar, or whole number or to fewer decimal places. For instance, the amount $56.129 is converted to $56.13 by rounding to the nearest cent.

When rounding numbers, locate the digit to the *right* of the point where rounding is to take place. If that digit is 5 or greater, round *up* by adding 1 to the digit at which rounding is to take place. If the digit to the right of the point where rounding is to take place is 4 or smaller, round *down* by leaving the digit at which rounding is to take place as it is.

Either drop all digits to the right of the rounded-off number (usually with whole numbers) or replace them with zeros (usually with decimals and cents), whichever is appropriate for the calculation.

In the following examples, the number to consider when determining rounding is circled.

EXAMPLES

Number	Round to	Answer (rounded to)
$25.17③	Nearest cent	$25.17
$25.17⑥	Nearest cent	$25.18
$1,497.⑨2	Nearest dollar	$1,498.00
$1,497.④9	Nearest dollar	$1,497.00
346.174⑧	3 decimal places	346.175
346.174③	3 decimal places	346.174
58.43⑥9%	2 decimal places	58.44%
58.43②6%	2 decimal places	58.43%

Mentally Multiplying by a Power of 10

To mentally multiply a whole number by 10 or a power of 10 (100, 1,000, and so on), add a zero to the multiplicand for each zero in the multiplier.

EXAMPLES

Multiplicand	×	multiplier	=	answer
385	×	10	=	3,850
4,672	×	100	=	467,200
219	×	1,000	=	219,000

To mentally multiply a decimal or a mixed number (whole number and decimal or dollar and cents amount) by 10 or a power of 10, move the decimal point one place to the right in the multiplicand for every zero in the multiplier. If there are not enough digits in the multiplicand (as in the second and last of the following examples) add zeros in the answer.

EXAMPLES

Multiplicand	×	multiplier	=	answer
57.289	×	10	=	572.89
$23.75	×	10	=	$237.50
.2056	×	100	=	20.56
4.3097	×	1,000	=	4,309.7
3.4	×	1,000	=	3,400

Mentally Dividing by a Power of 10

To mentally divide by 10 or a power of 10 (100, 1,000, and so on), move the decimal point one place to the left in the dividend for every zero in the divisor. If necessary (as in the last of the following examples), add zeros to the left of your answer. (*Note:* In a whole number, the decimal point is to the right of the last digit, even though it is not shown.)

EXAMPLES

Dividend	÷	divisor	=	answer
36.147	÷	10	=	3.6147
$479.80	÷	10	=	$47.98
25,689	÷	100	=	256.89
1,478.126	÷	1,000	=	1.478126
35.6	÷	1,000	=	.0356

Estimating Answers

An approximate answer to a calculation can be mentally estimated by rounding off all numbers in the problem and by then adding, subtracting, multiplying, or dividing the rounded-off numbers. When doing so, round off the numbers to amounts that are easy to work with, such as tens, hundreds, or thousands. This quick mental estimate will tell you the approximate amount or range of the actual answer.

EXAMPLES

Add		Mentally round off to this
385	=	400
208	=	200
475	=	500
+ 122	=	+ 100
1,190 (Actual answer)		1,200 (Mental estimate)

Subtract		Mentally round off to this
18,076	=	18,000
− 13,889	=	− 14,000
4,187 (Actual answer)		4,000 (Mental estimate)

Converting Fractions to Decimals

To convert a fraction to a decimal, divide the numerator (top number) by the denominator (bottom number). Carry the answer out to as many decimal places as are required for your purposes. Usually, two, three, or four decimal places (digits to the right of the decimal point) are sufficient. Round off the last digit of the decimal if necessary.

Zeros at the end of a decimal do not affect the decimal's value. Therefore, .5, .50, and .500 all have the same value.

EXAMPLE Determine the decimal equivalent of $\frac{3}{4}$.

Solution:

Decimal equivalent = numerator ÷ denominator

Decimal equivalent = 3 ÷ 4

Decimal equivalent = .75

The decimal equivalents of some commonly used fractions that appear in calculations in this book are shown below. You will find this information to be an especially handy reference when you perform calculations involving stock and bond quotations.

Fraction	=	Decimal equivalent	Fraction	=	Decimal equivalent
$\frac{1}{2}$	=	.5	$\frac{2}{8}$	=	.25
$\frac{1}{3}$	=	.3333	$\frac{3}{8}$	=	.375
$\frac{2}{3}$	=	.6667	$\frac{4}{8}$	=	.5
$\frac{1}{4}$	=	.25	$\frac{5}{8}$	=	.625
$\frac{2}{4}$	=	.5	$\frac{6}{8}$	=	.75
$\frac{3}{4}$	=	.75	$\frac{7}{8}$	=	.875
$\frac{1}{8}$	=	.125			

Multiplying Fractions

To multiply fractions, multiply the numerators (top numbers) by each other and multiply the denominators (bottom numbers) by each other. The answer is a new fraction that may be used as it is, reduced to lower terms, or converted to a decimal.

EXAMPLE Multiply $\frac{3}{8}$ by $\frac{2}{3}$.

Solution: Multiply first the numerators and then the denominators.

$$\frac{3}{8} \times \frac{2}{3} = \frac{3 \times 2}{8 \times 3} = \frac{6}{24}$$

The answer, $\frac{6}{24}$, can be reduced to $\frac{1}{4}$ by dividing both the numerator and denominator by 6, or it can be converted to its decimal equivalent, .25, by dividing the numerator by the denominator.

To multiply a whole number by a fraction, first convert the whole number to a fraction by making it the numerator of a fraction with 1 as the denominator. Then multiply as usual and perform whatever conversions are necessary.

EXAMPLE Multiply \$5.00 by $\frac{10}{365}$.

Solution: Convert \$5.00 to a fraction; multiply the numerators; and multiply the denominators.

$$\$5.00 \times \frac{10}{365} = \frac{\$5.00}{1} \times \frac{10}{365} = \frac{\$5.00 \times 10}{1 \times 365} = \frac{\$50.00}{365} = \$.1370$$

You will find this procedure for multiplying fractions to be particularly helpful when you perform interest calculations.

Calculating an Average

To calculate the average of a group of amounts, first total the amounts and then divide the total by the number of amounts.

EXAMPLE Charlie Jaynes bowled scores of 186, 207, and 171. Calculate his average for the three games.

Solution:

Step 1. Total the amounts:

$$\begin{array}{r} 186 \\ 207 \\ 171 \\ \hline 564 \end{array}$$

Step 2. Determine the average:

$$\begin{aligned} \text{Average} &= \text{total} \div \text{number of amounts} \\ \text{Average} &= 564 \div 3 \\ \text{Average} &= 188 \end{aligned}$$

Individual amounts are often compared with an average to determine if they represent an acceptable level of performance.

chapter 2

Using Percents

"It was announced today that city employees will receive a 7 percent pay raise." "According to the Department of Agriculture, this year's wheat crop should exceed last year's by 12 percent." "Joanie scored 98 percent on her algebra test—I'm so proud!" "Did you see the survey? Seventy-nine percent of the people in this town. . . ."

Percents—they're all around us in newscasts, government reports, and everyday conversation. They are used as a basis for making comparisons because they often provide a clearer and more accurate perspective than the amounts themselves would provide.

Since percent calculations are used in so many other types of computations, this section is presented early in the book. You will find it a helpful guide when you perform calculations which involve percents.

Understanding Percents

In any percent calculation there are three elements: the base, the rate, and the percentage. Their relations are expressed by the formula

$$\text{Percentage} = \text{base} \times \text{rate}$$

The *base* is the whole amount or original amount with which other amounts are compared. The *rate* is identified with the percent sign; it indicates the percent that an amount is of the base. The *percentage* is an amount calculated by multiplying the rate (percent) by the base. If any two elements of the percent formula are known, the third can be calculated.

It is usually easier to recognize the percent calculation to be made if elements of the formula are also identified with words that describe the calculation to be performed, as shown in each example and solution in this section. But before we perform any percent calculations, let's have a quick, 1-minute refresher course.

A percent shows the relation of a number to a base of 100. Thus, 25% can also be expressed as the decimal .25 or the fraction $\frac{25}{100}$. Likewise, 110% could be expressed as 1.10 or $\frac{110}{100}$.

To convert a percent to a decimal, remove the percent sign and move the decimal point two places to the left:

Percent	=	decimal
18%	=	.18
2.5%	=	.025
.5%	=	.005
110%	=	1.10

To convert a whole number or a decimal to a percent, move the decimal point two places to the right and add a percent sign:

Decimal	=	percent
.265	=	26.5%
.05	=	5%
3.51	=	351%
.025	=	2.5%

Remember that adding one or more zeros to the end of a decimal does not change the value of the decimal. Likewise, if a percent contains a decimal, a zero or zeros can be added to its end without changing its value. For instance, 25.7%, 25.70%, and 25.700% are identical in value.

Proving the Accuracy of Your Calculations

It's true that determining which one of the three percent calculations described below to use for a particular situation can be a little confusing. You can most likely identify the right procedure, though, by studying the examples provided, locating one which seems to be most like the calculation you want to perform, and by then plugging your own information into the formula.

To be certain your calculation is correct, you should then prove the accuracy of your work. To do so, remove one of the known amounts from the percent formula and insert the amount you have just calculated. Then do the calculation. If the answer you have calculated is correct, the result will be the known amount that you removed from the formula. This procedure is illustrated in each of the solutions in this section.

Calculating the Percent One Number Is of Another (Rate)

To calculate the percent one number is of another, the rate, divide the number being analyzed (the percentage) by the one to which it is being compared (the base). The percent should be carried out to the number of decimal places desired for your purposes. Usually, two decimal places are sufficient.

EXAMPLE Last year, Helen O'Keefe's total income was $32,000. Of that, $2,720 was from interest earned on investments. What percent of Helen's total earnings was from interest income?

Solution:

Rate	=	percentage	÷	base
Percent	=	interest income	÷	total income
Percent	=	$2,720	÷	$32,000
Percent	=		.085	
Percent	=		8.50%	

Proof: $32,000 × 8.50% = $2,720

Calculating an Amount (Percentage)

The percentage is an amount that is identified by its percent relationship to the base. If the rate (percent) is less than 100%, the percentage will be less than the base. If the rate is greater than 100%, the percentage will be greater than the base.

EXAMPLE 1 It has just been announced that all employees where Miguel Ramos works will receive a 8.5% pay increase, based on their current earnings. Currently, Miguel earns $32,500 per year. How much of a pay increase will he receive?

Solution:

Percentage	=	base	×	rate
Pay increase	=	current salary	×	percent of pay raise
Pay increase	=	$32,500	×	8.5%
Pay increase	=	$32,500	×	.085
Pay increase	=	$2,762.50		

Proof: $2,762.50 ÷ $32,500 = .085 (or 8.5%)

EXAMPLE 2 In March, sales at Madden Company were $55,000. In April, sales were 112% of March sales. What was the dollar amount of April sales?

Solution:

Percentage	=	base	×	rate
April sales	=	March sales	×	Percent April sales were of March sales
April sales	=	$55,000	×	112%
April sales	=	$55,000	×	1.12
April sales	=	$61,600		

Proof: $61,600 ÷ $55,000 = 1.12 (or 112%)

Calculating the Original Amount (Base)

The basé, or original amount, can be calculated by dividing the percentage by the rate.

EXAMPLE Keith Carlson boasted at a party that he had just received a pay raise of $1,500, which amounted to an 8% raise based on his previous salary. Calculate Keith's previous salary.

Solution:

Base	=	percentage	÷	rate
Previous salary	=	pay raise	÷	percent of raise
Previous salary	=	$1,500	÷	8%
Previous salary	=	$1,500	÷	.08
Previous salary	=	$18,750		

Proof: $18,750 × 8% = $1,500

Calculating the Percent of Increase

When the percent of increase from one amount to another is to be calculated, the original or oldest amount is the one with which comparison is made. Just follow these steps:

1. It will be obvious if there is an increase, since the most recent amount will be larger than the original (oldest) amount.
2. Deduct the original amount from the most recent amount to determine the amount of increase.
3. Divide the amount of increase by the original amount. The result is the percent of increase.

EXAMPLE Last week, Sharon Bileau purchased a box of Sweet 'N Tasty cereal for $1.55. This week, an identical box of Sweet 'N Tasty is priced at $1.84. Calculate the percent of increase in price from last week to this week.

Solution:

Step 1. Calculate the amount of increase.

Amount of increase	=	most recent amount	−	original amount
Amount of increase	=	$1.84	−	$1.55
Amount of increase	=		$.29	

Step 2. Calculate the percent of increase.

Percent of increase	=	amount of increase	÷	original amount
Percent of increase	=	$.29	÷	$1.55
Percent of increase	=		.1871	
Percent of increase	=		18.71%	

Proof: $1.55 × 18.71% = $.29

Calculating the Percent of Decrease

The procedure for calculating the percent of decrease from one amount to another is similar to that for calculating the percent of increase:

1. It will be obvious if there is a decrease, since the most recent amount will be smaller than the original (oldest) amount.
2. Deduct the most recent amount from the original amount to determine the amount of decrease.
3. Divide the amount of decrease by the original (oldest) amount. The result is the percent of decrease.

EXAMPLE Last year, Harvey Watland's salary was $35,000. This year, because of companywide pay cuts, Harvey's salary is $33,425. What percent did Harvey's salary decrease from last year to this year?

Solution:

Step 1. Calculate the amount of decrease.

Amount of decrease	=	original amount	−	most recent amount
Amount of decrease	=	$35,000	−	$33,425
Amount of decrease	=		$1,575	

Step 2. Calculate the percent of decrease.

Percent of decrease	=	amount of decrease	÷	original amount
Percent of decrease	=	$1,575	÷	$35,000
Percent of decrease	=		.045	
Percent of decrease	=		4.5%	

Proof: $35,000 × 4.5% = $1,575

FREQUENT PERCENT CALCULATIONS

As evidenced by the foregoing examples, percents are used in many types of calculations. Some of the most common percent calculations that occur in nearly everyone's life are identified in the following sections.

Calculating Sales Tax

The amount of sales tax that applies to a purchase is calculated by multiplying the sales tax rate by the purchase price of the goods. The sales tax amount is rounded off to the nearest cent. The total amount due is calculated by adding the sales tax to the price of the goods.

EXAMPLE Ann Peters purchased a vacuum cleaner for $389.99. There is a 5% sales tax in Ann's state. Calculate (*a*) the amount of sales tax and (*b*) the total amount due.

Solution: (*a*) Calculate the sales tax.

Sales tax	=	purchase price	×	sales tax rate
Sales tax	=	$389.99	×	5%
Sales tax	=	$389.99	×	.05
Sales tax	=		$19.50	

(*b*) Calculate the total amount due.

Total amount due = purchase price of goods + sales tax

Total amount due = $389.99 + $19.50

Total amount due = $409.49

The amount of sales tax calculated by this procedure may vary by a cent or two from that calculated by a sales clerk if the clerk uses a sales tax chart instead of multiplying the purchase price by the sales tax percent.

Calculating Sales Discounts

Merchants often offer sales during which the price of goods is reduced by a certain percent. The amount of the sales discount is calculated by multiplying the original price by the sales discount percent. The sales price is calculated by deducting the sales discount from the original price.

EXAMPLE Barb's Fashions is holding a storewide sale, with 20% off on all items in the store. Last week, before the sale began, Sally McCrutchen looked at a coat priced at $269.50. Calculate (*a*) the amount of discount that will apply to the coat and (*b*) the sales price of the coat.

Solution: (*a*) Calculate the sales discount.

Sales discount = original price × sales discount percent

Sales discount = $269.50 × 20%

Sales discount = $269.50 × .20

Sales discount = $53.90

(*b*) Calculate the sales price.

Sales price = original price − sales discount

Sales price = $269.50 − $53.90

Sales price = $215.60

Calculating a Down Payment

Often when goods are purchased on a layaway or installment plan, a down payment is required. The amount of down payment is calculated by multiplying the down payment percent by the price of the goods.

EXAMPLE Muller Appliance Center requires a 25% down payment on all appliances sold on the installment plan. Richard Rozen plans to buy a microwave oven priced at $459.75. How much down payment will he be required to make if he buys on the installment plan? Round the answer off to the nearest cent.

Solution:

Down payment	=	price of goods	×	down payment percent
Down payment	=	$459.75	×	25%
Down payment	=	$459.75	×	.25
Down payment	=		$114.94	

Calculating Tips

It is common practice to tip a waiter or waitress 15% of the bill, and tips are also commonly given to others such as cab drivers and bellboys. Since most people do not have a calculator handy at such times, or would feel it inappropriate to use one, the calculation must be made mentally. Mental calculation of a 15% tip can be performed as follows:

1. Start by calculating 10% of the total amount by moving the decimal point one place to the left. For instance, 10% of $38.80 is $3.88.
2. Round off the amount calculated in step 1 to an even number that is easy to work with. For instance, round off $3.88 to $4.00 or to $3.80 or $3.90.

3. If you desire to leave a 15% tip, it is $1\frac{1}{2}$ times the 10% amount calculated in steps 1 and 2. The tip can be calculated mentally by dividing the 10% estimate by 2 and then adding that amount to the 10% estimate. For instance, if the 10% estimate amounts to \$4.00, the 15% tip will be \$6.00 (\$4.00 ÷ 2 = \$2.00; \$4.00 + \$2.00 = \$6.00).

EXAMPLE Sandra Bellvue bought lunch for a number of friends. The bill came to \$59.70. Sandra wanted to leave as a tip approximately 15% of the bill. How much should she have left?

Solution:

Step 1. Mentally calculate 10% of the bill:

10% of \$59.70 is \$5.97.

Step 2. Round off the 10% amount:

Round off \$5.97 to \$6.00.

Step 3. Mentally calculate the 15% tip:

\$6.00 ÷ 2 = \$3.00

\$6.00 + \$3.00 = \$9.00

This procedure will help you estimate a tip of a certain percent. You can, of course, then raise or lower the amount given as a tip as you see fit.

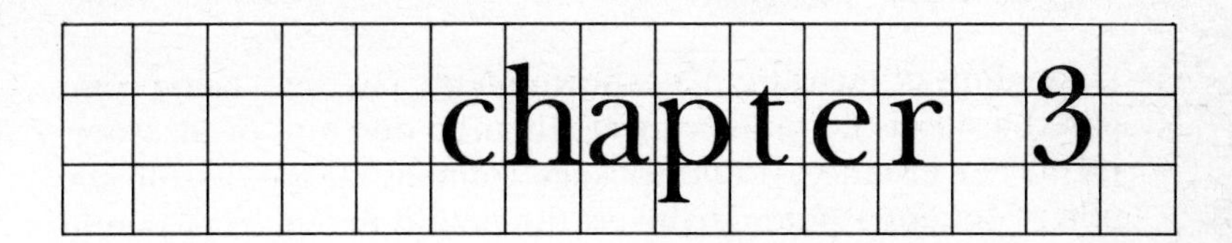

Household Calculations

One of the easiest ways to "earn" money is to make good use of the money that you already have. Often, you can do that by shopping intelligently, by renting various household articles instead of buying them, or, in some cases, by buying instead of renting frequently.

This section presents steps to follow in making various calculations that might arise in the day-to-day operation of a household. Applying these calculations to your daily life can help you become a more intelligent buyer, which will, in turn, save you money.

Comparing Prices of Different-Size Packages

The shelves of grocery stores are crammed with various containers of the same product bearing labels such as "regular," "super," "giant," and "jumbo." So is it better to buy three regular-size boxes instead of one giant-size box? How about buying one jumbo-size container instead of two super-size ones? It's all pretty confusing, but the answer can be easily determined by performing one simple mathematical calculation: division.

All containers have labels that show the number of ounces, pounds, gallons, or other unit of measurement of the products contained in them. To compare the prices of various sizes of the same product, proceed as follows:

1. If the units of measurement shown on the packages being compared are not the same, convert them to one unit of measurement. For instance, if one package contains 10 ounces and the other contains 1 pound, convert the pound to ounces (1 pound = 16 ounces).

2. Divide the price of each package by the number of grams, ounces, pounds, or other unit of measurement in the package. The result is the cost per unit of measurement. This division can be performed easily by using a hand-held calculator. Carry your answer out to at least four decimal places to get an accurate comparison.

3. The comparison can be extended. You can calculate the cost of buying the quantity contained in the larger-size package at the price per unit of the smaller-size package. Also, you can calculate the savings in buying one larger package or several smaller packages. These steps are illustrated in the following example and solution.

EXAMPLE A "regular" box of Austin detergent weighs 20 ounces and costs $1.03; a "jumbo" box of the detergent contains 10 pounds, 12 ounces and costs $6.59. Mary Jacobs wants to buy either the jumbo-size box or enough regular-size boxes to equal approximately the quantity in the jumbo-size box. Calculate (*a*) the cost per ounce of a regular-size package, (*b*) the cost per ounce of a jumbo-size package, (*c*) the cost of buying the jumbo-size quantity (10 pounds, 12 ounces) at the price per ounce of the regular-size package, and (*d*) the amount that Mary can save by buying the equivalent of the jumbo size in the most economical packaging.

Solution: (*a*) Calculate the cost per ounce of a regular-size package.

Cost per ounce = price ÷ number of ounces

Cost per ounce = $1.03 ÷ 20

Cost per ounce = $.0515 (or 5.15 cents)

(*b*) Calculate the cost per ounce of a jumbo-size package.

Step 1. Convert the package weight to ounces.

$$\text{Number of ounces} = \left(\begin{array}{c}\text{number}\\ \text{of pounds}\end{array} \times \begin{array}{c}\text{ounces}\\ \text{per pound}\end{array}\right) + \begin{array}{c}\text{additional}\\ \text{ounces}\end{array}$$

$$\text{Number of ounces} = (10 \times 16) + 12$$

$$\text{Number of ounces} = 160 + 12$$

$$\text{Number of ounces} = 172$$

Step 2. Calculate the cost per ounce.

$$\text{Cost per ounce} = \text{price} \div \text{number of ounces}$$

$$\text{Cost per ounce} = \$6.59 \div 172$$

$$\text{Cost per ounce} = \$.0383 \text{ (or 3.83 cents)}$$

(*c*) Calculate the cost of buying 172 ounces (jumbo-size quantity) at the regular-size package price per ounce.

$$\text{Cost} = \begin{array}{c}\text{regular-size package}\\ \text{price per ounce}\end{array} \times \begin{array}{c}\text{number}\\ \text{of ounces}\end{array}$$

$$\text{Cost} = \$.0515 \times 172$$

$$\text{Cost} = \$8.86$$

(*d*) Calculate the saving by buying 172 ounces (jumbo-size quantity) in the most economical packaging.

$$\text{Saving} = \begin{array}{c}\text{cost of 172 ounces in}\\ \text{regular-size package}\end{array} - \begin{array}{c}\text{cost of 172 ounces in}\\ \text{jumbo-size package}\end{array}$$

$$\text{Saving} = \$8.86 - \$6.59$$

$$\text{Saving} = \$2.27$$

Usually, the unit-of-measurement price of the larger packages is lower than that of the smaller ones. That is not always true, however, particularly when smaller packages are on sale. Also, it is not economical to buy a far larger quantity of an item than you need just to get a lower cost per unit, since the extra quantity will be wasted.

Buying in Quantity

Often, better prices per item can be obtained by buying in lots of two, three, or more than by buying just a single item. That is particularly true when special sales are offered. If the product is something you know you will need to buy again in the near future, and if it can be easily stored without spoiling, it is ordinarily wise to buy in money-saving quantities.

To determine the significance of buying in quantity, it is a good idea to calculate both the amount and percent of the saving, which can be done as follows:

1. Calculate the cost per item at the quantity price. To do so, divide the total cost by the number of items.
2. Calculate the saving per item by deducting the quantity price per item from the regular price per item.
3. Calculate the total saving by multiplying the saving per item by the number of items purchased.
4. To calculate the percent of saving, divide the saving per item (as calculated in step 2) by the regular cost per item.

EXAMPLE A bar of Roma soap is priced at 59¢ (cents). If bought in quantity, the price is three bars of soap for $1.50. Calculate (*a*) the total amount that will be saved by purchasing six bars of Roma soap at the quantity price instead of making six individual purchases and (*b*) the percent of saving that buying in quantity yields. Round your percent off at two decimal places.

Solution: (*a*) Calculate the total amount of saving by buying in quantity.

Step 1. Calculate the cost per item at the quantity price.

Quantity price per item = total cost ÷ number of items
Quantity price per item = $1.50 ÷ 3
Quantity price per item = $.50 (or 50¢)

Step 2. Calculate the savings per item.

Saving per item = regular price per item − quantity price per item
Saving per item = $.59 − $.50
Saving per item = $.09 (or 9¢)

Step 3. Calculate the total saving by buying at the quantity price.

Total saving = saving per item × number of items purchased

Total saving = $.09 × 6

Total saving = $.54 (or 54¢)

(*b*) Calculate the percent of the saving.

Percent of saving = saving per item ÷ regular price per item

Percent of saving = $.09 ÷ $.59

Percent of saving = .1525

Percent of saving = 15.25%

Even though the dollar amount of saving in this example is small, the percent saved is substantial. If similar quantity purchases can be made over a period of time, the shopper will be able to purchase many additional items with the money saved by using this technique.

Using Coupons

Coupons—magazines, Sunday newspaper supplements, and other publications are full of them. Most of them offer cents off on the purchase of a product. That's nice, but is it worth taking the time and effort to leaf through publications in search of a few coupons that might save you a dollar or two? The answer to that question becomes clearer if the percent of saving is analyzed.

To calculate the percent saved by utilizing a cents-off coupon, simply divide the value of the coupon by the regular price of the product.

EXAMPLE While reading a magazine, Mary Jo Helmke finds a coupon offering 50¢ off on the purchase of a 3-pound can of her favorite coffee. On her next grocery shopping trip, Mary Jo applies the coupon toward the purchase of the coffee, which has a regular price of $6.29. Calculate the percent Mary Jo saves by using the coupon.

Solution:

Percent of saving	=	coupon value	÷	regular price of product
Percent of saving	=	$.50	÷	$6.29
Percent of saving	=		.0795	
Percent of saving	=		7.95%	

The preceding calculation illustrates that use of a coupon can produce an excellent saving on purchases. Coupons must be used selectively, however. Use them only for the purchase of products that you need and would buy anyway. If you purchase items that you don't actually need, just to utilize a coupon, the result will be a waste of money rather than a saving.

Actively pursuing the use of coupons, manufacturers' rebates, and similar consumer deals can cut the cost of groceries, and often of other household products, substantially—perhaps by 20% or more.

Renting versus Buying

How many products can you name that are used perhaps only once or twice per year, if that often, but which you do have a definite need for when the time arises? Household articles that fall into this classification might include a carpet shampooer, punch bowl, silver service, rug stretcher, floor sander, and candelabra. About such products this question might be asked: "Is it financially better to rent the item or to buy it?"

Ultimately, only you can answer the question on the basis of your own perspectives. As an aid to an intelligent decision, though, a simple calculation can be made. This calculation will show the number of times you can rent the item before the total rental charges will equal the cost of buying the item. If you are able to estimate the number of times you will use the item per year, you can also calculate an estimated number of years it would take before the total rental charges would equal the purchase price. Here is how to proceed.

1. Determine these factors:

   a. The length of time for which you will need to use the item per occasion (such as 2 hours or 1 day)

   b. The number of times per year that you will need to use the item

   c. The cost of renting the item per occasion of use

   d. The cost of buying the item

2. To determine the number of times you could rent the item before the total rental charges would equal the cost of buying it, divide the cost of buying the item by the rental charge per occasion of use.

3. To determine the number of years you could rent the item before the total rental charges would equal the purchase price, take these steps:

   a. Calculate the total rental cost per year by multiplying the rental charge per occasion by the number of times the item will be used per year.

   b. Divide the purchase price of the item by the total rental charge per year (as calculated in step *a*).

EXAMPLE 1 Vivian Hubers shampoos her carpets twice a year. She needs a carpet shampooer for 2 days each time she shampoos. Vivian can rent a shampooer from a local hardware store for $20 per day ($40 per rental: $20 per day × 2 days). On the other hand, she can buy an adequate shampooer for $200. Calculate (*a*) the number of times Vivian can rent the shampooer before the total rental charges equal the cost of buying a shampooer and (*b*) the number of years she can rent the shampooer before the total rental charges equal the cost of buying one.

Solution: (*a*) Calculate the number of times the shampooer can be rented before rental charges equal cost of buying one.

Number of rentals	=	cost of buying item	÷	rental charge per use
Number of rentals	=	\$200	÷	\$40
Number of rentals	=		5	

(*b*) Calculate the number of years the shampooer can be rented before rental charges equal the cost of buying.

Step 1. Calculate the rental cost per year.

Rental cost per year	=	rental charge per use	×	number of times used per year
Rental cost per year	=	\$40	×	2
Rental cost per year	=		\$80	

Step 2. Calculate the number of years before the total rental charges equal the cost of buying.

Number of years	=	cost of buying item	÷	rental cost per year
Number of years	=	\$200	÷	\$80
Number of years	=		2.5 (or $2\frac{1}{2}$ years)	

EXAMPLE 2 Ken Hickman cuts and splits his own wood for use in his fireplace. Each year, for \$30, he rents a wood splitter for one day, which is sufficient time to split enough firewood to meet an entire season's needs. Ken can purchase a wood splitter for \$1,500. Calculate the number of years Ken can rent a wood splitter before the cost of renting equals the cost of buying one.

Solution:

Number of years	=	cost of buying item	÷	rental cost per year
Number of years	=	\$1,500	÷	\$30
Number of years	=		50	

In the example of the carpet shampooer, it can be seen that purchase may be a wise financial move, since the shampooer will pay for itself in $2\frac{1}{2}$ years. In the wood splitter example, it would take 50 years for the splitter to pay for itself in saved rental charges and therefore purchase may not be wise.

Of course, many other factors, such as the convenience and pride of owning the item, need to be considered. Also, if an item is purchased, there are apt to be costs for maintenance and repairs. If the item is rented, those costs are ordinarily borne by the business providing the rental service. Another thought is that if you buy an item such as a wood splitter, you may be able to rent it out to others and recoup part or all of the purchase price.

Of course, in many cases, there is a better solution to either renting or buying a seldom-used item: borrow it from a friend!

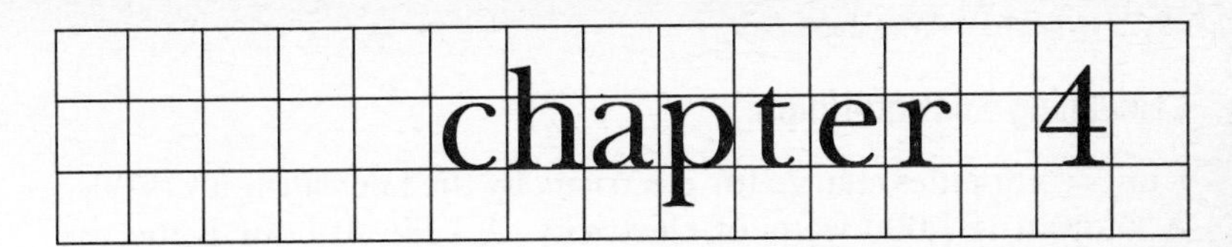

Energy and Utilities Calculations

Energy conservation is a term that has become a part of everyone's vocabulary in the past few years. Undoubtedly it will increase in both use and importance of application.

This section describes how to calculate energy and utilities costs and provides information that will help you analyze your utility bills and the cost of using various forms of energy. Also presented are steps to take in analyzing the payback on the purchase of energy-efficient devices and the savings that can be reaped by applying energy conservation techniques.

CALCULATING ELECTRICITY COSTS

"Turn off the lights—you're wasting electricity." How many times were you told that as a child or have you said it to your spouse or a child of your own? Certainly it is true that needless use of electricity is wasteful. But just how much does it cost to leave that light burning in an unoccupied room? Well, this section will answer that question and other questions about electricity costs.

Calculating Kilowatt-Hours

Utility companies charge for electricity by the kilowatt-hour (kWh). A kilowatt is 1,000 watts of electricity. A kilowatt-hour is the use of 1,000 watts for an hour's time.

To calculate the number of kilowatt-hours of electricity used by a light bulb, appliance, or other electrical product, follow this procedure:

1. Calculate the number of watt-hours used by multiplying the product's wattage (as stated on the light bulb or product) by the number of hours of use.
2. Calculate the kilowatt-hours of electricity used by dividing the watt-hours (as calculated in step 1) by 1,000, which is the number of watts in a kilowatt.

EXAMPLE The Rubel family use their 400-watt color television set an average of 180 hours per month (6 hours per day × 30 days). Calculate the number of kilowatt-hours of electricity used by the television set in a typical month.

Solution:

Step 1. Calculate the watt-hours.

$$\text{Watt-hours} = \text{wattage} \times \text{no. of hours used}$$
$$\text{Watt-hours} = 400 \times 180$$
$$\text{Watt-hours} = 72{,}000$$

Step 2. Calculate the kilowatt-hours used.

$$\text{Kilowatt-hours} = \text{watt-hours} \div \text{no. of watts in a kilowatt}$$
$$\text{Kilowatt-hours} = 72{,}000 \div 1{,}000$$
$$\text{Kilowatt-hours} = 72$$

Electric motors used in large appliances, electric tools, and other products are rated in terms of horsepower. One horsepower equals 746 watts.The number of kilowatt-hours used by an electric motor is calculated as follows:

1. Calculate the number of watts used by the electric motor by multiplying the number of horsepower stated on the motor by 746, the watts equal to 1 horsepower.
2. Calculate the number of watt-hours used by multiplying the motor's wattage (as calculated in step 1) by the number of hours the motor is used.
3. Calculate the kilowatt-hours of electricity used by dividing the watt-hours (as calculated in step 2) by 1,000, the number of watts in a kilowatt.

EXAMPLE Dan Sullivan uses his table saw, which has a 2-horsepower electric motor, an average of 15 hours per month. Calculate the number of kilowatt-hours of electricity used to operate the saw in a typical month.

Solution:

Step 1. Calculate the watts used by the electric motor.

$$\text{Watts} = \text{horsepower} \times \text{no. of watts per horsepower}$$
$$\text{Watts} = 2 \times 746$$
$$\text{Watts} = 1{,}492$$

Step 2. Calculate the watt-hours.

$$\text{Watt-hours} = \text{wattage} \times \text{no. of hours used}$$
$$\text{Watt-hours} = 1{,}492 \times 15$$
$$\text{Watt-hours} = 22{,}380$$

Step 3. Calculate the kilowatt-hours.

$$\text{Kilowatt-hours} = \text{watt-hours} \div \text{no. of watts per horsepower}$$
$$\text{Kilowatt-hours} = 22{,}380 \div 1{,}000$$
$$\text{Kilowatt-hours} = 22.38$$

Reading the Electricity Bill

The number of kilowatt-hours of electricity used in a billing period (usually a month) is often stated on the utility bill. If it is not stated, it can be calculated by deducting the preceding electrical meter reading from the current meter reading.

EXAMPLE On a recent utility bill received by the Atkens family, the current meter reading shows 79,278 kilowatt-hours and the preceding meter reading shows 78,193 kilowatt-hours. Calculate the number of kilowatt-hours of electricity used during the billing period.

Solution:

$$\text{Kilowatt-hours used} = \begin{matrix}\text{current}\\ \text{meter reading}\end{matrix} - \begin{matrix}\text{previous}\\ \text{meter reading}\end{matrix}$$

$$\text{Kilowatt-hours used} = 79{,}278 - 78{,}193$$

$$\text{Kilowatt-hours used} = 1{,}085$$

Calculating the Cost per Kilowatt-Hour

The cost per kilowatt-hour may be stated on your utility bill, or it can be determined by telephoning the power company. Often, a sliding scale such as the following one is used.

ELECTRICITY RATES

Kilowatt-hours used	Cost per kilowatt-hour
First 400	$.080
Next 200	$.075
Balance	$.065

With this scale, the consumer pays the highest rate for the first set number of kilowatt-hours of electricity used (400), pays a lower rate for the next set number of kilowatt-hours used (200), and so on.

EXAMPLE The Atkens family used 1,085 kilowatt-hours of electricity last month. Use the following scale to calculate the cost of electricity used.

Kilowatt-hours used	Cost per kilowatt-hour
First 350	$.075
Next 250	$.068
Balance	$.056

Solution:

Kilowatt-hours used	×	cost per kilowatt-hour	=	total electricity cost
350	×	$.075	=	$26.25
250	×	$.068	=	17.00
485	×	$.056	=	27.16
1,085				$70.41

Calculating Average Cost per Kilowatt-Hour

Often, for purposes of calculating the cost of electricity used, it is easier to use the average cost per kilowatt-hour rather than attempt to use a sliding scale as described above. To calculate the average cost per kilowatt-hour, divide the total electricity cost by the kilowatt-hours used, as shown on a recent electricity bill.

EXAMPLE Last month the Atkens family used 1,085 kilowatt-hours of electricity at a cost of $70.41. Calculate the average cost per kilowatt-hour. Round your answer off at four decimal places.

Solution:

$$\text{Average cost per kilowatt-hour} = \text{total electricity cost} \div \text{kilowatts used}$$

Average cost per kilowatt-hour = $70.41 ÷ 1,085
Average cost per kilowatt-hour = $.0649 (or 6.49¢)

Calculating the Cost of Operating Electrical Products

The preceding calculations pertaining to electricity are interesting, but what are they good for? Well, they are necessary to enable us to make an important calculation: the cost of operating various electrical products. To calculate the cost of operating an electrical product, proceed as follows:

1. Calculate the number of kilowatt-hours used by the product over the time period being analyzed (as described on page 32).
2. Determine your average cost per kilowatt-hour as described in the preceding section.
3. Multiply the number of kilowatt-hours used (as calculated in step 1) by the average cost per kilowatt-hour (as calculated in step 2).

EXAMPLE 1 The Coleman family estimate that they use their 300-watt color television set 1,750 hours per year. Mr. Coleman calculated the average cost per kilowatt-hour of electricity to be $.0685. Calculate the electricity cost of operating the television set for a year.

Solution:

Step 1. Calculate the watt-hours used.

Watt-hours = wattage × no. of hours used
Watt-hours = 300 × 1,750
Watt-hours = 525,000

Step 2. Calculate the kilowatt-hours used.

Kilowatt-hours = watt-hours ÷ no. of watts in a kilowatt
Kilowatt-hours = 525,000 ÷ 1,000
Kilowatt-hours = 525

Step 3. Calculate the electricity cost.

$$\text{Electricity cost} = \text{kilowatt-hours used} \times \text{cost per kilowatt-hour}$$

$$\text{Electricity cost} = 525 \times \$.0685$$

$$\text{Electricity cost} = \$35.96$$

EXAMPLE 2 Dale Stevens estimates that, through carelessness, the equivalent of one 75-watt light bulb is left burning in an unoccupied room in his home for 720 hours each year (2 hours per day × 360 days). Mr. Stevens has calculated the average cost per kilowatt-hour of electricity to be $.072. Calculate the cost of electricity wasted because the Stevens family fail to turn off unneeded lights.

Solution:

Step 1. Calculate the watt-hours used.

$$\text{Watt-hours} = \text{wattage} \times \text{no. of hours used}$$

$$\text{Watt-hours} = 75 \times 720$$

$$\text{Watt-hours} = 54{,}000$$

Step 2. Calculate the kilowatt-hours used.

$$\text{Kilowatt-hours} = \text{watt-hours} \div \text{no. of watts in a kilowatt}$$

$$\text{Kilowatt-hours} = 54{,}000 \div 1{,}000$$

$$\text{Kilowatt-hours} = 54$$

Step 3. Calculate the electricity cost.

$$\text{Electricity cost} = \text{kilowatt-hours used} \times \text{cost per kilowatt-hour}$$

$$\text{Electricity cost} = 54 \times \$.072$$

$$\text{Electricity cost} = \$3.89$$

Even though $3.89 may not seem like much, it will add up over a period of years, particularly if the same pattern of carelessness extends to all other electrical products in the home.

One of the most valuable ways to use calculations of electricity costs is in comparing products, such as stoves, refrigerators, clothes dryers, and water heaters, that use large amounts of electricity. The wattage or horsepower of each of these products, as well as the average cost of operation per year, is usually stated on the product's label. The cost figure is based on averages, so it is wise to make your own calculations by using your own average hours of use of the product and your local electricity rates.

CALCULATING NATURAL GAS COSTS

If you have a natural gas furnace, water heater, range, air conditioner, and clothes dryer, or just some of those appliances, the annual cost of the natural gas you use may well exceed all of your other utility costs put together. If that is your plight, you will find it helpful to understand how natural gas bills are calculated. Also, you will find it virtually essential to be able to make calculations pertaining to decreasing natural gas consumption and cutting utility bills, as explained in this section.

Reading the Natural Gas Bill

Natural gas is measured by the cubic foot. Your natural gas bill, however, probably states meter readings in terms of *hundred cubic feet.* If it does, a meter reading of 2,467 actually represents 2,467 *hundred* cubic feet (or 246,700 cubic feet) of gas. On many gas bills, the abbreviation CCF is used to identify hundred cubic feet. (The first C is the abbreviation for 100; the last two letters, CF, stand for cubic feet.) The term *therm,* which is sometimes used, stands for 100 cubic feet.

Most gas bills identify the preceding meter reading, present meter reading, and CCF of gas used during the billing period. If you want to make your own calculation of the amount of gas used during a billing period, subtract the preceding meter reading from the present meter reading. Also, if you want to determine the amount of gas used during a day, week, or other time period, you can take your own meter readings and make a similar calculation.

EXAMPLE At 9:00 p.m., January 29, Fred Samuels read his natural gas meter, which showed 4,318. At 9:00 p.m., January 30, Fred again read his gas meter; this time it showed 4,322. Calculate the CCF (hundred cubic feet) of gas consumed during the 24-hour period.

Solution:

CCF used = present meter reading − previous meter reading

CCF used = 4,322 − 4,318

CCF used = 4 (or 400 cubic feet)

Calculating the Cost per CCF

Some natural gas companies use a sliding scale like the one shown below to calculate the cost of natural gas used. Remember that CCF stands for 100 cubic feet.

NATURAL GAS RATES

Gas used	Cost per CCF
First 5 CCF	$.838
Next 45 CCF	$.569
Next 450 CCF	$.550
Balance	$.539

By this scale, the consumer pays the highest rate for the first set number of CCF of gas used (5), pays a lower rate for the next set number of CCF used (45), and so on.

EXAMPLE During the last month, the Chalmers family used 78 CCF of natural gas. Using the following scale to calculate the cost of gas the family used.

Gas used	Cost per CCF
First 4 CCF	$.745
Next 30 CCF	$.625
Next 100 CCF	$.575
Balance	$.516

Solution:

CCF used	×	Cost per CCF	=	Total gas cost
4	×	$.745	=	$ 2.98
30	×	$.625	=	18.75
44	×	$.575	=	25.30
78				$47.03

Many consumers feel that use of a sliding scale of rates penalizes those who have taken steps to become energy-efficient and decrease their gas consumption. That is because higher rates are paid by everyone for the first CCF used, but those who use huge quantities of gas, and who are perhaps wasteful or inefficient, buy most of their gas at the lower rates.

Therefore, the trend by natural gas companies is to charge every customer a monthly meter use charge, such as $10, and charge all customers the same rate, say, $.555 per CCF, for all the gas used. The procedure for calculating the cost of gas used under that billing arrangement is similar to the one shown above: multiply the number of CCF of gas used by the cost per CCF, and add to that the monthly meter charge.

EXAMPLE Last month, the Zwick family used 86 CCF of natural gas. The utility company charges all users $.555 per CCF and assesses a $10 monthly meter charge. Calculate the Zwick's utility bill for the month.

Solution:

Step 1. Calculate the cost of the gas.

Cost for gas = CCF used × cost per CCF

Cost for gas = 86 × $.555

Cost for gas = $47.73

Step 2. Calculate the total utility bill.

Total utility bill = cost for gas + monthly meter charge

Total utility bill = $47.73 + $10.00

Total utility bill = $57.73

To determine your natural gas rates and the method used in calculating your gas bill, contact your natural gas supplier.

Calculating the Average Cost per CCF

For the purpose of comparing the costs of operating various natural-gas-burning appliances, the average cost per CCF is often adequate. To calculate the average cost per CCF, simply divide your total natural gas bill by the number of CCF used. This method can be used regardless of whether your natural gas supplier uses a sliding scale or a flat rate plus meter charge as a billing method.

EXAMPLE The Wilcox family's most recent natural gas bill shows that 182 CCF of gas was used. The total bill was $113.49. Calculate the average cost per CCF. Round your answer off at three decimal places.

Solution:

Average cost per CCF = total gas bill ÷ no. of CCF used
Average cost per CCF = $113.49 ÷ 182
Average cost per CCF = $.624 (or 62.4¢)

Cutting Your Natural Gas Costs

There are three basic ways, as described below, to reduce your natural gas expenditures. Some of these cost nothing or very little and can give you an immediate return for your efforts. Others involve the expenditure of substantial amounts of money, but the long-term payback may still make them very economical. Taking advantage of all of the ways may allow you to cut your total natural gas costs by as much as 25% to 40% or more, depending upon your current use of energy-efficient appliances, devices, and techniques.

It is often difficult for an individual to determine precisely where and how much natural gas is being wasted because of improper home insulation, inefficient appliances, and so on. Therefore, it is advisable to invest in a *home energy audit,* in which an energy efficiency consultant will study all aspects of your home and natural gas use and make recommendations for cutting natural gas costs.

Since an energy audit may take 3 to 4 hours and the data may be processed and analyzed by computer, the cost of the audit may be $200 or more. In many states, however, the commerce commission or some similar administrative agency requires natural gas suppliers to conduct energy audits for a certain percent of their customers annually. In order to comply with those regulations, natural gas suppliers usually find it necessary to charge their customers only a fraction of the actual cost of an energy audit, say, $15, to entice enough customers to have a home energy audit made. The natural gas supplier absorbs the remaining cost of the energy audit. Therefore, if your natural gas supplier does offer an energy audit at a similar cost, it is a valuable service at a bargain price that you should consider utilizing.

Efficient Use of Appliances and Devices. The easiest and least expensive way to cut natural gas costs is to make efficient use of your gas-burning appliances and devices. For instance, simply setting your gas water heater thermostat in the range of 120 degrees, instead of 150 to 160 degrees or higher, will save a very substantial amount. Setting your furnace thermostat lower during the winter heating season and your air conditioner thermostat higher during the summer cooling season will also produce big savings. Another saver is to run your clothes dryer only with a full load—never a partial load. Still another is to avoid opening the oven door repeatedly when cooking, thus releasing the heat from the oven and requiring more natural gas to reheat it. Yet another is to decrease the time spent in the shower, thus cutting your water heater costs.

As you survey your daily activities, you will undoubtedly uncover many similar situations in which you can decrease your natural gas use. All you need to invest is a little time and effort.

Sealing Your Home. If you have them, a natural gas furnace or air conditioner are most likely the largest users of gas in your home. Therefore, if you can seal your home so there is less escape of heat in the winter and cool air in the summer, you will be able to permanently decrease your gas consumption. Often, a few dollars worth of caulking applied around windows and door jambs and used to seal other cracks will provide a huge payback. Likewise, attic and sidewall insulation can provide large permanent savings. Installing storm doors and thermally insulated draperies and using similar techniques can also cut gas costs.

To determine the number of years it will take to pay back the cost of insulation, caulking, and so on, divide the cost of the improvement by the estimated saving per year. To calculate the percent that the annual saving represents, divide the estimated annual saving by the cost of the improvement.

EXAMPLE Dave and Julie Ackerman hired a consultant to perform an energy audit of their home. The consultant estimated that the Ackermans will save $208 per year in natural gas heating and air-conditioning costs if they spend $650 to have their attic insulated properly. Calculate (*a*) the number of years it will take for the saving to pay for the cost of the improvement and (*b*) the annual percent of return on their investment for insulation that the saving represents.

Solution: (*a*) Calculate the number of years to pay back the cost.

No. of years = cost of improvement ÷ annual savings

No. of years = $650 ÷ $208

No. of years = 3.125 (or $3\frac{1}{8}$ years)

(*b*) Calculate the annual percent of return on investment.

Annual percent of return = annual savings ÷ cost of improvement

Annual percent of return = $208 ÷ $650

Annual percent of return = .32

Annual percent of return = 32%

In the preceding example, it can be seen that the investment for insulation would be wise, since it will pay for itself in a very short time period and will continue to provide savings in the years thereafter. The 32% rate of return on their investment is very high—certainly greater than the Ackermans could realize if they put their money in a savings account or took some other investment opportunity. Then too, if they install the insulation themselves rather than hire someone to do it, they can increase their savings and percent of return far beyond that shown here.

If the total savings over an extended time period, say, 10 years, is considered, the percent of return on the Ackerman's investment is phenomenal. In fact, if the Ackermans do not have the cash available to pay for the insulation, borrowing the money to take advantage of the huge potential savings will be well worthwhile. The percent of annual savings will greatly exceed the interest rate charged on the loan.

Buying Energy-Efficient Appliances. Most gas appliances have energy-guide labels that show their average costs of operation. Since the figures are averages, your actual cost of operating the appliance may be higher or lower, depending upon your method of operation, frequency of use, and other factors that vary from person to person.

Still, energy-guide labels and average cost figures can provide information that is valuable in comparing one manufacturer's model to another's when you are contemplating a purchase. Often, you will find that appliances that are higher in efficiency are also higher in cost because they are more expensive to manufacture. Your goal, then, should be to calculate the amount of potential annual saving by buying the most efficient model and calculate the number of years it will take for the accumulated savings to pay the extra cost of the more highly efficient but more expensive product. The average energy cost figures shown on the product's energy-guide label will provide an adequate comparison, but if possible, make the comparison by using your own actual gas consumption figures as indicated by an energy audit. Here are the steps in making your calculation:

1. Determine the difference in cost between the products being compared by deducting the price of the least expensive product from that of the most expensive product.
2. Calculate the annual saving by buying the most efficient model.

Deduct the annual cost of operation of the most efficient product from that of the least efficient product.

3. Calculate the number of years it will take for the annual savings to pay back the extra cost of buying the more efficient, more expensive model. To do so, divide the difference in cost of the two models (as calculated in step 1) by the annual savings (as calculated in step 2).
4. To calculate the annual percent of return on the investment by buying the more efficient, more expensive product, divide the annual saving (as calculated in step 2) by the difference in cost of the products being compared (as calculated in step 1).

EXAMPLE The Fong's furnace is worn out and must be replaced. Through comparative shopping, they have selected two furnace models for analysis. A conventional furnace is 64% efficient; its installed cost is $900; and its average energy cost is $645 per year. An energy-saver model is 95% efficient; its installed cost is $2,200; and its average energy cost is $504 per year. Calculate (*a*) the difference in cost of the two models, (*b*) the annual saving by buying the most efficient model, (*c*) the number of years it will take for the annual energy saving to pay back the additional cost of the more expensive model, and (*d*) the annual return on investment—the percent that the annual saving is of the extra cost of buying the more expensive model.

Solution: (*a*) Calculate the difference in cost of the two models.

$$\text{Cost difference} = \text{cost of energy-saver furnace} - \text{cost of conventional furnace}$$

$$\text{Cost difference} = \$2{,}200 - \$900$$

$$\text{Cost difference} = \$1{,}300$$

(*b*) Calculate the annual saving in gas cost by buying the more efficient model.

$$\text{Annual saving} = \text{annual cost of operating conventional furnace} - \text{annual cost of operating energy-saver furnace}$$

Annual saving	=	$645	−	$504
Annual saving	=		$141	

(*c*) Calculate the number of years of payback by buying the more expensive model.

No. of years of payback	=	cost difference	÷	annual savings
No. of years of payback	=	$1,300	÷	$141
No. of years of payback	=	9.22 (about 9 years, 3 months)		

(*d*) Calculate the annual return on investment that the annual saving represents.

Percent of return	=	annual saving	÷	cost difference
Percent of return	=	$141	÷	$1,300
Percent of return	=		.1085	
Percent of return	=		10.85%	

When making this type of analysis, keep in mind that many experts believe the cost of natural gas will increase substantially in future years. Therefore, the amount and percent of saving may be much higher than this analysis predicts. Also, the saving will continue for many years beyond the payback period needed to recoup the extra cost of the more expensive model. All in all, it would appear from the example that to buy the more efficient, but more expensive, model would be an excellent decision.

CALCULATING WATER COSTS

Do you know how much the water to take a 10-minute shower costs you? Most likely your answer is, "I don't have the foggiest idea." Well, the first step in finding out is to determine the cost per gallon of water furnished by your local utility company. Then, we'll determine how many gallons of water that shower takes, and we'll soon have the answer.

Utility companies charge for water by either the gallon or by the cubic foot, as described below.

By-the-Gallon Billings

The number of gallons of water used during a billing period is usually stated on the utility bill. If it is not, it can be determined by deducting the current meter reading from the preceding meter reading, both of which are ordinarily shown on the bill.

EXAMPLE In Marlon Stenzel's community, water is charged for by the gallon. Marlon's utility bill shows a current water meter reading of 92,016 and a preceding water meter reading of 85,238. Calculate the number of gallons used during the billing period.

Solution:

Gallons used = current meter reading − previous meter reading

Gallons used = 92,016 − 85,238

Gallons used = 6,778

Water Cost per Gallon. The cost per gallon of water is either printed on the utility bill or can be determined by making a telephone call to the utility company. Often, a sliding scale like the following one is used for water charges.

UTILITY COMPANY WATER RATES

Gallons used	Cost per gallon
First 1,850	$.0045*
Next 5,500	$.0038
Next 30,000	$.0034
Next 60,000	$.0027
Balance	$.0021

*$.0045 = .45¢, or slightly less than one-half cent per gallon

By this scale, the consumer pays the highest rate for the first set number of gallons used (1,850), pays a lower rate for the next set number of gallons (5,500), and so on.

EXAMPLE The Andrews family used 8,210 gallons of water last month. Use the following water rate scale to calculate the cost of the water used.

Gallons used	Cost per gallon
First 2,000	$.0048
Next 5,000	$.0040
Next 25,000	$.0034

Solution:

Gallons used	×	cost per gallon	=	total water cost
2,000	×	$.0048	=	$ 9.60
5,000	×	$.0040	=	20.00
1,210	×	$.0034	=	4.11
8,210				$33.71

Calculating Average Cost per Gallon. To calculate the cost of water used for a specific household task, it is usually easier to use an average cost per gallon rather than apply a sliding scale. To calculate the average cost per gallon, simply divide the total water cost for the billing period by the number of gallons used.

EXAMPLE Last month, the Andrews family used 8,210 gallons of water at a cost of $33.71. Calculate the average cost per gallon. Round your answer off at four decimals.

Solution:

Average cost per gallon	=	total water cost	÷	no. of gallons used
Average cost per gallon	=	\$33.71	÷	8,210
Average cost per gallon	=		\$.0041 (or .41¢)	

Billing by the Cubic Foot

Some utility companies charge for water by the number of cubic feet used. First, the number of cubic feet of water used is either determined or read from the utility bill. It is determined by deducting the preceding meter reading from the current meter reading, exactly as described in the preceding section. The cost per cubic foot can be found by referring to a sliding water rate scale or by calculating the average cost per cubic foot used. The procedure for making the calculation is given in the preceding section.

For ease of calculating the cost of water use, the number of cubic feet of water consumed can be converted to gallons. The conversion is made by multiplying the number of cubic feet used by 7.48, the approximate number of gallons of water in a cubic foot. The average cost per gallon of water can be calculated by dividing the total water bill by the number of gallons used.

EXAMPLE In Tony Rizzo's community, water is charged for by the cubic foot. Tony's utility bill shows a current water meter reading of 24,030 and a preceding water meter reading of 22,750. The cost of the water is \$25.59. (*a*) Calculate the number of cubic feet of water used. (*b*) Convert the number of cubic feet of water used to gallons. (*c*) Determine the average cost per gallon. Round your answer off at five decimal places.

Solution: (*a*) Calculate the cubic feet of water used.

Cubic feet used	=	current meter reading	−	previous meter reading
Cubic feet used	=	24,030	−	22,750
Cubic feet used	=		1,280	

(*b*) Convert cubic feet to gallons.

Gallons = cubic feet used × gallons per cubic foot

Gallons = 1,280 × 7.48

Gallons = 9,574.40

(*c*) Calculate the average cost per gallon.

Average cost per gallon = total cost ÷ no. of gallons used

Average cost per gallon = $25.59 ÷ 9,574.40

Average cost per gallon = $.00267 (or .267¢)

Calculating an Appliance's Water Use Cost

The number of gallons of water used by a washing machine, dishwasher, or other appliance ordinarily can be found in the literature which accompanies the appliance. To calculate the cost of water, multiply the number of gallons used by the cost per gallon.

EXAMPLE The new shower head which Tim Johnsom purchased uses 8 gallons of water per minute. Each of the six members of the Johnson family takes a daily shower which averages 5 minutes. Mr. Johnson has calculated the average cost of water to be $.0036 per gallon. Calculate (*a*) the water cost of one 5-minute shower and (*b*) the total water cost of showers for the Johnson family in a year.

Solution: (*a*) Calculate the water cost of one shower.

Cost of one shower = (gallons per minute × minutes used) × cost per gallon

Cost of one shower = (8 × 5) × $.0036

Cost of one shower = 40 × $.0036

Cost of one shower = $.144 (or 14.4¢)

(*b*) Calculate the year's water cost.

Step 1. Calculate the number of gallons used in a day.

$$\text{Gallons used per day} = \left(\frac{\text{gallons}}{\text{per minute}} \times \frac{\text{minutes}}{\text{per person}}\right) \times \frac{\text{number of}}{\text{persons}}$$

$$\text{Gallons used per day} = (8 \times 5) \times 6$$

$$\text{Gallons used per day} = 40 \times 6$$

$$\text{Gallons used per day} = 240$$

Step 2. Calculate the number of gallons used in a year.

$$\text{Gallons used in a year} = \frac{\text{gallons used}}{\text{per day}} \times \frac{\text{number of days}}{\text{in a year}}$$

$$\text{Gallons used in a year} = 240 \times 365$$

$$\text{Gallons used in a year} = 87{,}600$$

Step 3. Calculate the water cost of showers in a year.

$$\text{Water cost} = \text{gallons used in a year} \times \text{cost per gallon}$$

$$\text{Water cost} = 87{,}600 \times \$.0036$$

$$\text{Water cost} = \$315.36$$

In the preceding example, if the Johnsons would limit their showers to 4 minutes (a 20% decrease in time) each, they could cut the annual water bill by approximately $63.07 (20% of the year's $315.36 total).

Calculating the Payback on Water-Saving Devices

Buying water-efficient appliances and other water-saving devices can often substantially decrease your water bill. The cost of the new appliances and devices should be weighed against the savings obtained from them. This can be done by multiplying the number of gallons of water saved per year by the average cost per gallon to determine the annual saving. Then the cost of the water-saving device is divided by the saving per year to determine the number of years it will take for the saving in water use to pay for the item.

The annual percent of return that the saving represents can be calculated by dividing the annual saving by the cost of the item.

EXAMPLE The Johnson's shower head uses 8 gallons of water per minute. Each of the six members of the Johnson family takes a 5-minute shower daily. The average cost per gallon of water is $.0036. Therefore, in a year, the Johnsons run the shower 10,950 minutes (6 showers daily × 5 minutes each × 365 days = 10,950 minutes) and use 87,600 gallons of water at a total water cost of $315.36 for their showers.

For a cost of $11.98, Mr. Johnson can buy a water-saving shower head that uses only 3 gallons of water per minute. Calculate (*a*) the gallons of water that can be saved per year with the water-saving shower head, (*b*) the dollars in water costs that can be saved per year, (*c*) the number of years it will take for the savings to pay for the water-saving device, and (*d*) the annual percent of return that the annual saving is of the cost of the device.

Solution: (*a*) Calculate the gallons saved per year.

Step 1. Calculate the gallons saved per minute.

Gallons saved per minute	=	gallons per minute, old shower head	−	gallons per minute, new shower head
Gallons saved per minute	=	8	−	3
Gallons saved per minute	=		5	

Step 2. Calculate the gallons saved per year.

Gallons saved per year	=	gallons saved per minute	×	no. of minutes used per year
Gallons saved per year	=	5	×	10,950
Gallons saved per year	=		54,750	

(*b*) Calculate the dollars saved per year.

Dollars saved per year	=	gallons saved	×	cost per gallon
Dollars saved per year	=	54,750	×	$.0036
Dollars saved per year	=		$197.10	

(*c*) Calculate the number of years of payback.

Years of payback	=	cost of device	÷	saving per year
Years of payback	=	$11.98	÷	$197.10
Years of payback	=	.06 year (about 22 days!)		

(*d*) Calculate the annual percent of return on the item's cost.

Annual percent of return = annual saving ÷ cost of item

Annual percent of return = $197.10 ÷ $11.98

Annual percent of return = 16.4524

Annual percent of return = 1,645.24%

This example dramatically illustrates the point that some water-saving devices are well worth the money. Obviously, not all water-saving devices will yield such an immediate payback. The example is quite realistic for the family of six, however, since some regular shower heads do use up to 8 gallons of water per minute and some water-saving shower heads do purport to use 3 gallons or less per minute.

CALCULATING SEWER COSTS

In essence, utility companies ordinarily set their sewer rates as a percent of the customer's water bill, since most of the water used is discharged into the sewer lines. Therefore, if you are able to decrease your water usage, and therefore your water bill, you will most likely decrease your sewer costs as well.

The amount of saving in the sewer bill that will result from decreasing water use can be calculated as follows:

1. Calculate the percent your sewer cost is of the water cost. Do that by dividing the sewer cost by water cost on a recent utility bill.
2. Calculate the dollars saved on your water bill through the use of some water-saving device or technique (as described in the preceding section).
3. Multiply the percent your sewer cost is of your water cost (as calculated in step 1) by the saving in your water cost (as calculated in step 2). The result is the approximate amount of money your sewer bill will decrease as a result of decreasing your water usage.

EXAMPLE Last month the Dixon's water bill was $40.24 and their sewer bill was $19.47. Mr. Dixon just purchased several water-saving devices which he estimates will reduce water use and save $7 in water costs per month. Calculate (*a*) the percent that the Dixon's sewer cost is of their water cost and (*b*) the estimated saving in sewer costs that will result from the decrease in water use.

Solution: (*a*) Calculate the percent the sewer cost is of the water cost.

Percent = sewer cost ÷ water cost

Percent = $19.47 ÷ $40.24

Percent = .4838

Percent = 48.38%

(*b*) Calculate the saving in sewer costs

Saving in sewer costs = savings in water costs × percent sewer costs are of water costs

Saving in sewer costs = $7.00 × 48.38%

Saving in sewer costs = $3.39

When analyzing the effect of buying water-saving devices or of using water-saving techniques, therefore, the resulting saving in sewer costs should also be included.

THE REAL COST OF PAYING A UTILITY BILL LATE

So you were a few days late in paying your utility bill and you were assessed a penalty of a few dollars. So what? Before we casually dismiss the cost of a utility bill late payment penalty, let's take a closer look.

On some utility bills, the penalty is labeled as *penalty*. On others, two amounts are shown: the *net amount due* and the *gross amount due*. The net amount due is the actual amount of your bill, which is the amount you are to pay if you make the payment by the due date. The gross amount due, which is larger than the net amount due, includes the amount of the utility bill plus the penalty. If you do not pay by the due date, you owe the gross amount.

The penalty for late payment varies from company to company, but it may be as high as 10%. Or is it? Actually, the *real cost* of making a late payment and being assessed a penalty is ordinarily beyond the 10% range. Far beyond it. That is because interest rates charged for loans are always stated in terms of the *annual percentage rate* (APR). The APR is the interest rate charged for the loan of money over a year's time.

In effect, when you pay your utility bill late, the utility company is loaning you the money until you make your payment and the penalty assessed can be considered your cost of interest on the loan. Therefore, to determine its real cost, we must view a utility bill late payment penalty in terms of the annual percentage rate, just as we would any other loan.

The formula for calculating the annual percentage rate of a utility bill late payment penalty is as follows:

$$\text{Rate} = \frac{\text{amount of penalty}}{\text{amount due} \times \text{time}}$$

The amount due is the net amount before the penalty is assessed. The time is a fraction with the number of days the payment is made beyond the due date as the numerator (top number) and the number of days in a year, 365, as the denominator (bottom number).

EXAMPLE Neil Kiminsky's utility bill showed a net amount due of $67.46 if the bill was paid by November 15 and a gross amount due of $70.84 if the bill was paid after the 15th. Neil paid the bill 5 days late, on November 20. Calculate the annual percentage rate that the late payment penalty amounted to.

Solution:

Step 1. Calculate the penalty amount.

$$\text{Penalty} = \text{gross amount due} - \text{net amount due}$$

$$\text{Penalty} = \$70.84 - \$67.46$$

$$\text{Penalty} = \$3.38$$

Step 2. Calculate the annual percentage rate of the late payment penalty.

$$\text{Annual percentage rate} = \frac{\text{amount of penalty}}{\text{amount due} \times \text{time}}$$

$$\text{Annual percentage rate} = \frac{\$3.38}{\$67.46 \times \frac{5}{365}}$$

$$\text{Annual percentage rate} = \frac{\$3.38}{\frac{\$67.46}{1} \times \frac{5}{365}}$$

$$\text{Annual Percentage rate} = \frac{\$3.38}{\frac{\$337.30}{365}}$$

$$\text{Annual percentage rate} = \frac{\$3.38}{\$.9241^{*}}$$

$$\text{Annual percentage rate} = 3.6576$$

$$\text{Annual percentage rate} = 365.76\%$$

Thus it can be said that, in the preceding example, Mr. Kiminsky could "earn" 365.76% on his money (the $67.46) simply by paying his utility bill on time. In a single month, this does not amount to many dollars saved, but over a period of several years, the amount would be substantial.

BUYING FIREWOOD

As more and more people have turned to wood burners and energy-efficient fireplaces to augment their home's heating, the purchase of firewood has become more common. Firewood is usually sold by the cord, bundle, or pickup load.

A cord is an exact measure; it consists of a stack of wood 8 feet long, 4 feet wide, and 4 feet high. It contains 128 cubic feet, which is calculated by multiplying the length by the width by the height ($8 \times 4 \times 4 = 128$).

*$337.30 ÷ 365 = $.9241.

In cities particularly, firewood is often sold by the bundle, which can vary in size from one seller to another. To estimate the percent that a bundle is of a cord and determine if the price for the bundle is reasonable, follow this procedure:

1. Estimate the bundle's length, width, and height in feet.
2. Multiply the bundle's estimated length by its width and then by its height. The result is the estimated cubic feet in the bundle.
3. Divide the bundle's estimated cubic feet by the number of cubic feet in a cord, 128. The result is the percent of a cord that the bundle represents.
4. To determine if the price of the bundle is in line with the price charged for a cord, divide the bundle price by the percent the bundle is of a cord (as calculated in step 3). The result is the price that would be charged for a cord of wood at the rate being charged for the bundle. This price can then be compared with that ordinarily charged for a cord of wood in your area.

EXAMPLE A firewood supplier is selling firewood at a city shopping center parking lot for $35 for a bundle approximately 3 feet long, 3 feet high, and 2 feet wide. Several recent advertisements in a local newspaper offer firewood at $150 a cord. Calculate (*a*) the number of cubic feet in the bundle offered by the firewood supplier, (*b*) the percent the bundle is of a cord, and (*c*) the estimated per cord price for the bundle.

Solution: (*a*) Calculate the estimated cubic feet in the bundle.

Estimated cubic feet in bundle = length × width × height

Estimated cubic feet in bundle = 3 × 2 × 3

Estimated cubic feet in bundle = 18

(*b*) Calculate the percent the bundle is of a cord.

Percent of a cord = cubic feet in bundle ÷ cubic feet in a cord

Percent of a cord = 18 ÷ 128

Percent of a cord = .1406

Percent of a cord = 14.06%

(*c*) Calculate the per cord price for the bundle.

Per cord price	=	price of bundle	÷	percent bundle is of a cord
Per cord price	=	$35.00	÷	14.06%
Per cord price	=	$35.00	÷	.1406
Per cord price	=	$248.93		

It is reasonable to expect to pay more than the normal per cord price when buying small quantities of firewood. Exactly how much more will depend upon the convenience provided, the quality of the wood, the quantity you want to purchase, and personal factors.

It is difficult to estimate the percent of a cord contained in a pickup load. That is because pickup boxes differ in size and the wood is usually thrown into the box casually, with indeterminable amounts of air space between the logs. Therefore, you may not be able to make a calculation like that shown above until the wood is delivered, unloaded, and stacked.

CALCULATING MILES PER GALLON

Not too many years ago, when the price of gasoline was under $.50 per gallon, many auto owners weren't overly concerned about keeping track of the miles per gallon obtained by their vehicles. Now, however, with the price of gasoline fluctuating over the range of $1.15 to $1.40 per gallon, and with periodic threats of the price going even higher, almost everyone is concerned about vehicular performance.

To calculate the miles per gallon obtained by your vehicle, and the cost of gasoline per mile driven, follow this procedure:

1. When you fill your vehicle's tank with gas, record the miles shown on the odometer (mileage indicator). Do not record the gallons of gas required to fill the tank or its cost.
2. Next time you fill the tank with gas, record the number of gallons required, the cost of filling the tank, and the miles shown on the odometer.

3. Determine the number of miles driven by deducting the original odometer reading (as recorded in step 1) from the current odometer reading (as recorded in step 2).
4. Divide the number of miles driven (as calculated in step 3) by the number of gallons of gas used. The result is the miles per gallon.
5. To calculate the gasoline cost per mile driven, divide the cost of refilling the tank (as recorded in step 2) by the number of miles driven (as calculated in step 3).

If you wish to calculate your miles per gallon and gasoline cost per mile for a trip which will require several refills, start by following the procedure shown in step 1. Then, each time you buy gas, record the number of gallons and the cost. When you reach your destination, follow the procedure described in steps 2 to 5.

EXAMPLE Jerry Becker filled his car's tank with gas and recorded the odometer reading of 34,176.4. After returning home from a trip to another city, he had the tank refilled, which took 18.6 gallons and cost $24.16. He then recorded the odometer reading, which was 34,576.3. Calculate (*a*) the miles per gallon and (*b*) the cost of gasoline per mile driven.

Solution: (*a*) Calculate the miles per gallon.

Step 1. Calculate the miles driven.

Miles driven = current odometer reading − original odometer reading

Miles driven = 34,576.3 − 34,176.4

Miles driven = 399.9

Step 2. Calculate the miles per gallon.

Miles per gallon = miles driven ÷ number of gallons used

Miles per gallon = 399.9 ÷ 18.6

Miles per gallon = 21.5

(*b*) Calculate the gasoline cost per mile driven.

Gas cost per mile = total gas cost ÷ miles driven
Gas cost per mile = $24.16 ÷ 399.9
Gas cost per mile = $.0604 (or 6.04¢)

Analyzing Vehicle Speed and Gas Savings

Each vehicle has a speed range in which it operates most economically. The speed range can be determined by keeping accurate records of miles per gallon when the driving is over similar routes and under similar conditions. The saving that can be obtained by driving at the most efficient speed can then be calculated by multiplying the number of gallons saved by the average cost per gallon.

EXAMPLE Leonard Blum drove 1,800 miles on a cross-country trip. On the way to his destination, he drove 62 miles per hour (mph) and averaged 21.5 miles per gallon. On his return trip, also 1,800 miles, Leonard drove 55 mph and averaged 25.1 miles per gallon. The average cost of gas was $1.249 per gallon. Calculate the amount of money Leonard saved in gas costs by driving 55 mph on the 1,800-mile return trip.

Solution:

Step 1. Calculate the number of gallons used at 62 mph.

Gallons used = miles driven ÷ miles per gallon
Gallons used = 1,800 ÷ 21.5
Gallons used = 83.72

Step 2. Calculate the number of gallons used at 55 mph.

Gallons used = miles driven ÷ miles per gallon
Gallons used = 1,800 ÷ 25.1
Gallons used = 71.71

Step 3. Calculate the number of gallons saved at 55 mph.

Gallons saved = gallons used at 62 mph − gallons used at 55 mph
Gallons saved = 83.72 − 71.71
Gallons saved = 12.01

Step 4. Calculate the saving from driving 55 mph.

Saving = gallons saved × average cost per gallon

Saving = 12.01 × \$1.249

Saving = \$15.00

Keep in mind that the saving calculated above is for only 1,800 miles of travel. If Leonard were to have 20,000 miles of similar highway travel per year, his annual saving would be approximately \$166.00.

Mileage per gallon can also be increased by keeping tires properly inflated, getting regular engine tuneups and avoiding sudden starts and stops and by not leaving the engine idle needlessly for long periods of time.

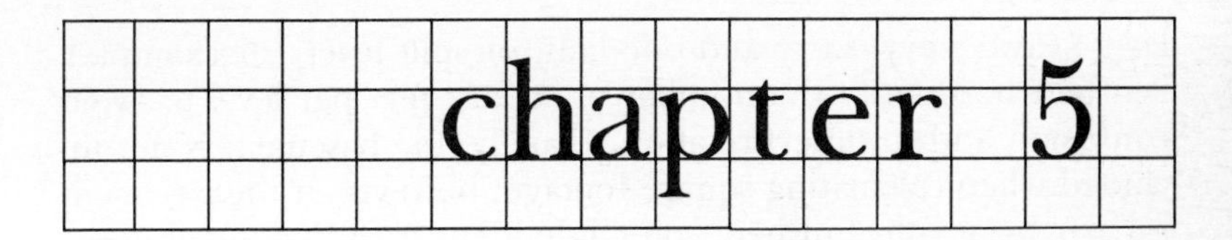

Building and Decorating Calculations

When you are contemplating some decorating or building, such as painting, wallpapering, or carpeting a room or pouring a concrete patio, you can save time and often money by making your own calculations before you contact a supplier or contractor. You will be able to determine the amount of materials necessary to do the job, estimate your costs, and make your decision on whether or not to proceed with your plans.

Calculating the Number of Square Feet in Your Home

One of the primary factors included in the description of a home by real estate salespeople, contractors, and home owners is the number of square feet in the home. This statistic is also of value when you are comparing homes for possible purchase or attempting to determine the value of your present home by comparing it with other homes that have sold in your area recently.

When you calculate the number of square feet in your home, use the *outside* measurements. That is, measure around the outside of your house or obtain the outside wall measurements of your condominium or cooperative. If your home has more than one level (a two-story, story and one-half, or split level, for example), it is best to identify the number of square feet per level to avoid confusion and misinterpretation. Usually, the basement is not included when calculating square footage; however, if the basement is partially or fully finished, you might identify the basement measurements separately. An attached garage is not included in calculating the square footage of a home, but its measurements and square footage can be shown separately.

Follow this procedure to calculate the number of square feet in your home:

1. Obtain a blueprint of your home or draw a sketch which resembles the exterior walls of your home.
2. Determine the exterior wall measurements from your blueprint or by measuring each wall. On your sketch, label each wall with its measurements.
3. If your home is square or rectangular, the number of square feet in it is calculated by multiplying its width by its length.
4. If your home is not square or rectangular but instead has one or more setoffs (additions which jut out), proceed as follows:
   - *a.* Measure each wall and label your sketch.
   - *b.* Draw dashed lines on your sketch to divide the home into a series of squares or rectangles.
   - *c.* Label all measurements on each of the squares or rectangles of step *b*.
   - *d.* Calculate the square feet in each of the squares and rectangles by multiplying the width by the length.
   - *e.* Add together the number of square feet calculated for each part of the home. The result is the total number of square feet.
5. Follow the above procedures to calculate the number of square feet in each level of your home.

EXAMPLE The Quinns made a drawing like the one in Fig. 5-1 of their ranch-style home with two setoffs and identified all outside wall measurements. Calculate the number of square feet in the home.

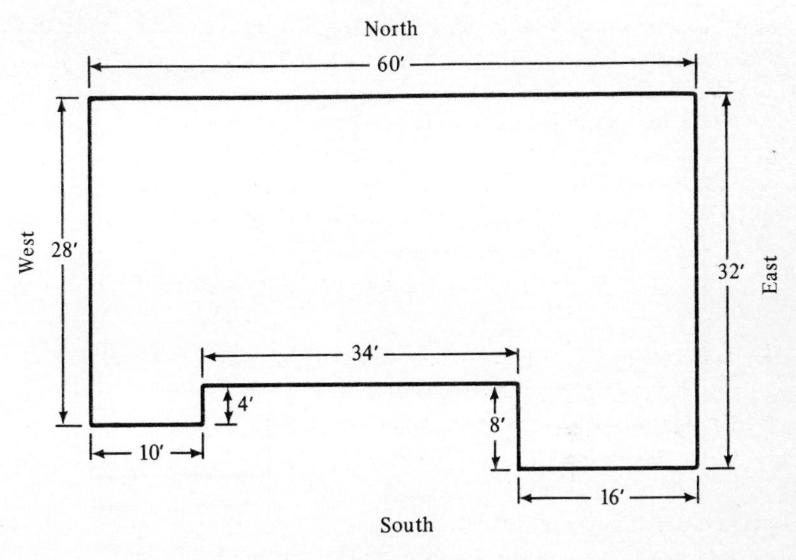

FIG. 5-1

Solution:

Step 1. Draw dashed lines to divide the sketch into a series of rectangles. (*Note:* Each "part" of the home is labeled with a letter for purposes of identification.)

Step 2. Label all measurements on each of the rectangular "parts" of the home. See Fig. 5-2.

a. It can be seen from the original sketch that part A is 4 feet × 10 feet.

b. It can be seen from the original sketch that part B is 8 feet × 16 feet.

c. It can be seen from the north wall measurement of the original sketch that part C is 60 feet long. The width of part C along the west wall is calculated to be 24 feet by deducting the 4-foot setoff of part A from the west wall's 28 feet.

The width of part C along the east wall is calculated to be 24 feet by deducting the 8-foot setoff of part B from the east wall's 32 feet.

Therefore, it can be seen that part C is a rectangle measuring 24 feet by 60 feet.

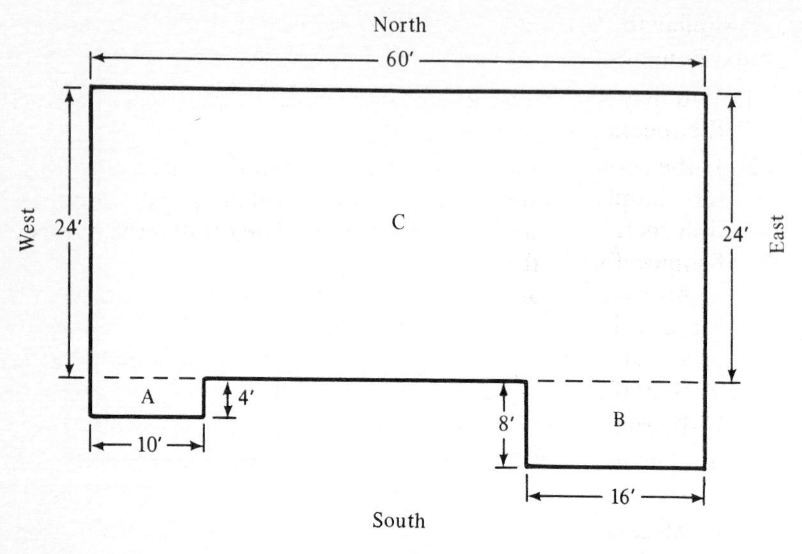

FIG. 5-2

Step 3. Calculate the square feet in the home.

Calculate the square feet of each part of the sketch by multiplying the part's width by its length. Add the square footage of all parts together to determine the total square feet in the home.

Part	Width, in feet	×	length, in feet	=	square feet in part
A	4	×	10	=	40
B	8	×	16	=	128
C	24	×	60	=	1,440
Total square feet in home				=	1,608

Calculating the Square Feet of Floor Space

Calculating the number of square feet of floor space in a room is often the first step in determining how many floor tiles or how many containers of varnish or wax will be needed to cover the area. The procedure for making this calculation, briefly described below, is similar to that used in determining the number of square feet in a home as described in the preceding section.

1. You may find it helpful to draw a sketch of the room and label the measurement on each wall.

2. If the room is square or rectangular, without any insets or offsets, simply measure the length of two adjoining walls and multiply their length, in feet, by each other. The result is the number of square feet in the room.

An *inset* is a portion of another room which extends into a room and thereby decreases its floor space. An *offset* is a portion of a room which juts out beyond a wall, as an adjoining closet does, and thereby adds floor space to the room.

3. If the room contains one or more insets, follow this procedure:

a. Calculate the entire room's square footage as though there were no insets, as described in step 2.

b. Measure the width and length of the floor space occupied by the inset.

c. Calculate the number of square feet that an inset occupies by multiplying its width by its length, in feet.

d. From the total square footage calculated in step *a,* deduct the square footage of the inset. The result is the number of square feet in the room.

4. If the room contains one or more offsets, follow this procedure:

a. Calculate the square footage of the room, as described in step 2 (and perhaps also step 3).

b. Measure the width and length of floor space occupied by the offset.

c. Calculate the number of square feet occupied by the offset by multiplying the offset's width by its length, in feet.

d. Add the square footage of the offset to the square footage of the rest of the room. The result is the total amount of floor space in the room.

EXAMPLE Dorothy Jewett wants to calculate the number of square feet in her bedroom. The room is 15 feet long and 12 feet wide. There is an inset measuring 2 feet by 4 feet and an offset measuring 3 feet by 8 feet. As an aid, she has drawn a sketch of the room like the one shown in Fig. 5-3 and labeled all of the walls' measurements. Calculate the number of square feet of floor space in the room.

Solution:

Step 1. Calculate the square feet in the room as if there were no insets or offsets.

$$\text{Square feet in room without insets or offsets} = \text{room's width} \times \text{room's length}$$

$$\text{Square feet in room without insets or offsets} = 12 \text{ feet} \times 15 \text{ feet}$$

$$\text{Square feet in room without insets or offsets} = 180$$

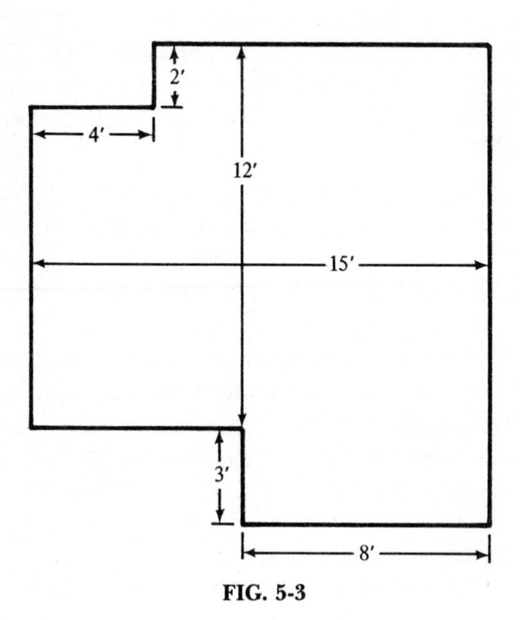

FIG. 5-3

Step 2. Calculate the square feet in the inset.

Square feet in inset = inset's width × inset's length

Square feet in inset = 2 feet × 4 feet

Square feet in inset = 8

Step 3. Calculate the square feet in the offset.

Square feet in offset = offset's width × offset's length

Square feet in offset = 3 feet × 8 feet

Square feet in offset = 24

Step 4. Calculate the square feet of floor space in room.

Square feet in room without inset or offset	180
Deduct square feet in inset	− 8
Add square feet in offset	+ 24
Square feet of floor space in the room	196

Calculating the Cost of Carpeting

Carpet is sold by the square yard, which measures 3 feet on a side and thus contains 9 square feet.

One might assume that the number of square yards of carpet needed to cover a floor is determined simply by finding the number of square feet in the room (as described in the preceding section) and then dividing that number by 9, the number of square feet in a square yard. That is often not the case, however, because carpet is not poured on the floor like water to spread out to cover every nook and cranny.

Carpet ordinarily comes in rolls which measure 12 feet wide. Therefore, if your room is less than 12 feet in width, you will have to buy the extra carpet that must be trimmed off—carpet that you must, nevertheless, pay for. Another solution might be to run the carpet the other direction on the floor, but then there will be a splice in the middle of the room and the carpet pattern might look awkward.

If your room is more than 12 feet wide, however, the only practical solution may be to run the carpet crosswise and splice it in the least conspicuous place. To determine the cost of carpeting a room, follow this procedure:

1. Determine in which direction the carpet should be laid to cover your floor. You may find it helpful to draw a sketch of the floor as an aid.
2. Calculate the length and width of carpet that must be purchased. Include any extra carpet that must be trimmed off.
3. Multiply the length by the width, in feet, of the carpet that must be purchased. This will yield the number of square feet of carpet to be bought.
4. Divide the number of square feet of carpet to be bought (as calculated in step 3) by 9, the number of square feet in a square yard. The result is the square yards of carpet to be purchased.
5. To calculate the cost of the carpet, multiply the number of square yards (as calculated in step 4) by the cost per yard.

EXAMPLE 1 Carpeting a room that is 12 feet wide.

A room that measures 12 feet by 20 feet is to be carpeted. The cost of the carpet is $10 per square yard. Calculate (*a*) the number of square yards of carpet needed and (*b*) the cost of the carpet.

Solution: (*a*) Calculate the square yards of carpet needed.

Step 1. Calculate the square feet of carpet needed.

Square feet of carpet needed = width, in feet × length, in feet

Square feet of carpet needed = 12 × 20

Square feet of carpet needed = 240

Step 2. Calculate the square yards of carpet needed.

Square yards of carpet needed = square feet of carpet needed ÷ square feet in a square yard

Square yards of carpet needed = 240 ÷ 9

Square yards of carpet needed = 26.67 (or $26\frac{2}{3}$)

(*b*) Calculate the cost of the carpet.

$$\text{Cost of carpet} = \begin{matrix}\text{square yards of}\\ \text{carpet needed}\end{matrix} \times \begin{matrix}\text{cost per}\\ \text{square yard}\end{matrix}$$

$$\text{Cost of carpet} = 26.67 \times \$10$$

$$\text{Cost of carpet} = \$266.70$$

EXAMPLE 2 Carpeting a room that is less than 12 feet wide.

A room measures 11 feet by 15 feet. The cost of carpet is $13 per square yard. Calculate (*a*) the number of square yards of carpet needed and (*b*) the cost of the carpet.

Solution: (*a*) Calculate the square yards of carpet needed.

Step 1. Calculate the square feet of carpet needed.

Even though the room is only 11 feet wide, a piece of carpet measuring 12 feet wide must still be purchased. The extra 1 foot of width will be trimmed off. Therefore, a piece of carpet 12 feet by 15 feet must be purchased. This piece of carpet contains 180 square feet as calculated by following the procedure shown in Example 1 (12 feet × 15 feet = 180 square feet).

Step 2. Calculate the square yards of carpet needed.

By following the procedure shown in Example 1, we find that 20 square yards of carpet is needed (180 square feet ÷ 9 square feet per yard = 20 square yards).

(*b*) Calculate the cost of the carpet.

By following the procedure shown in Example 1, we find that the carpet will cost $260 (20 square yards × $13 per square yard = $260).

EXAMPLE 3 Carpeting a room that is more than 12 feet wide.

A room measures 16 feet by 22 feet. The cost of carpet is $18 per square yard. Calculate (*a*) the number of square yards of carpet needed and (*b*) the cost of the carpet.

Solution: (*a*) Calculate square yards of carpet needed.

Step 1. Calculate the square feet of carpet needed.

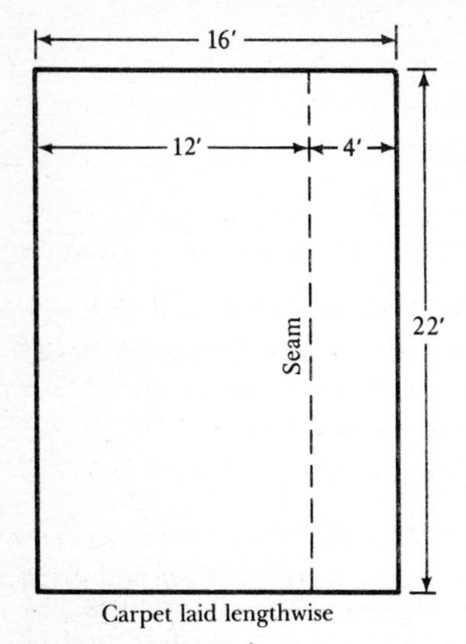

Carpet laid lengthwise

Fig. 5-4

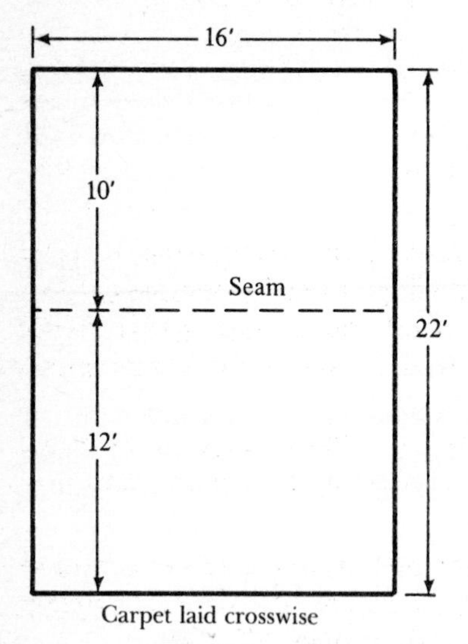

Carpet laid crosswise

Fig. 5-5

Since both the length and width of this room are greater than the 12-foot width of a roll of carpet, two pieces of carpet must be laid side by side and spliced together. This can be accomplished by laying the carpet either lengthwise or crosswise in the room.

If the carpet is laid lengthwise (see Fig. 5-4), one roll will cover a 12 foot by 22 foot area. Then another piece of carpet, measuring 12 feet by 22 feet, will be needed to cover the remaining 4 feet by 22 feet of floor space. In this case, a piece of carpet measuring 8 feet by 22 feet (costing you about $350) will have to be trimmed off. Since that is a great deal of waste, it may be best to lay the carpet crosswise in the room.

If the carpet is laid crosswise in the room one piece of carpet will cover an area 12 feet by 16 feet. That will leave a 10 foot by 16 foot area of floor space to cover. You will need to purchase a piece of carpet measuring 12 feet by 16 feet to cover the area. That will result in a piece of carpet measuring only 2 feet by 16 feet that will need to be trimmed off, which is about $285 *less* wasted carpet than if the carpet were laid lengthwise.

The two pieces of carpet, each measuring 12 feet by 16 feet, will come from a single roll, so you will actually purchase a piece of carpet measuring 12 feet by 32 feet. That piece of carpet will contain 384 square feet as calculated by following the procedure shown in Example 1 (12 feet × 32 feet = 384 square feet).

Step 2. Calculate the square yards of carpet needed.

By following the procedure shown in Example 1, we find that 42.67 (or $42\frac{2}{3}$) square yards of carpet is needed (384 square feet ÷ 9 square feet per yard = 42.67 square yards).

(*b*) Calculate the cost of the carpet.

By following the procedure shown in Example 1, we find that the carpet will cost $768.06 (42.67 square yards × $18 per square yard = $768.06).

In addition to buying the carpet, you may also need to buy a pad for underlayment beneath the carpet. That is true if the carpet you are buying has a jute back; a pad is not ordinarily needed if the carpet has rubber backing. Carpet pads are sold by the square yard, and the calculations are identical with those shown above for carpet. Also, the cost of installing the carpet by a carpet layer is additional, and installation is usually charged for by the square yard. Therefore, if the cost of the carpet is $15 per square yard, the pad is $4 per square yard, and the installation fee is $3 per square yard, your *installed* cost is $22 ($15 + $4 + $3) per square yard.

Linoleum and other floor coverings also are ordinarily sold by the square yard and are commonly available in rolls of 12-foot and 6-foot widths. Calculations for these floor coverings are similar to those shown in this section.

Calculating Gallons of Paint

Let's assume that you decide to repaint the living room of your home. How many gallons of paint will you need—one, two, three, more? Buying too little paint will send you scurrying back to the store, probably in paint-stained clothes, for more. Buying too much paint will be an expensive waste.

To calculate the amount of paint needed to cover a room, an entire home, or some other surface, and its cost, follow this procedure:

1. If all the walls are of the same height, measure the height of just one wall. If one or more walls are of a different height than the other walls, measure their heights as well.

2. Measure each wall's length. If opposite walls are of the same length, you need measure only one of them.
3. State your measurements in feet. For instance, 8 feet, 6 inches equals $8\frac{1}{2}$ feet, and so on.
4. Calculate the number of square feet in each wall by multiplying the wall's height by its length.
5. Calculate the total square feet of wall surface in the room (or other area) by adding the areas of the walls calculated in step 4.
6. Measure the height and width of all windows, doors, and other wall areas that will *not* be painted. State your measurements in feet.
7. Calculate the square feet of each area that will not be painted by multiplying its height by its width.
8. Calculate the total square feet of surface not to be painted by adding the individual areas calculated in step 7.
9. Calculate the number of square feet to be painted by deducting the square feet of wall areas not to be painted (calculated in step 8), from the total square feet of surface in the room (calculated in step 5).
10. Determine the number of gallons of paint needed by dividing the square feet of surface to be painted (calculated in step 9) by the number of square feet that a gallon of paint will cover. The number of square feet a gallon of paint will cover is shown on the paint container; it is usually 400 to 500 square feet. If two coats of paint will be needed, double your answer.

    If the number of gallons of paint needed is slightly less than a certain number of full gallons (such as 3.75 gallons), round the number up to the next full gallon—say, 3.75 gallons up to 4. Usually, this will be more economical than buying fewer gallons plus several quarts of paint. Also, it will provide a margin for error and a little paint for touch-up later on.

If the number of gallons needed is slightly more than a certain number of full gallons—say, 3.15 gallons, it is usually wise to round the number down to the next full gallon (as by rounding 3.15 gallons down to 3) and then buy an additional quart of paint. (*Note:* Four quarts equal one gallon.)

11. To calculate the cost of the paint needed, multiply the number of gallons needed by the cost per gallon and multiply the number of quarts needed by the cost per quart. Then add the two amounts together.

EXAMPLE Julie Holmberg plans to paint her living room with one coat of paint. Each wall is 9 feet high. The north and south walls are each 20 feet, 3 inches long ($20\frac{1}{4}$ or 20.25 feet), and the east and west walls are each 14 feet, 6 inches long ($14\frac{1}{2}$ or 14.5 feet). The room has one doorway, measuring 7 feet by $3\frac{1}{2}$ feet, one picture window, measuring $4\frac{1}{2}$ feet by 7 feet, and two other windows, each measuring 3 feet by 4 feet. A gallon of paint covers 450 square feet and costs $9.50; a quart covers 112.5 square feet and costs $2.75. Calculate (*a*) the square feet of wall space to be painted, (*b*) the number of gallons of paint needed, and (*c*) the cost of the paint.

Solution: (*a*) Calculate the square feet of wall space to be painted.

Step 1. Calculate the total square feet of wall space in the room.

Wall	Height, in feet	×	width, in feet	=	square feet in part
North	9	×	20.25	=	182.25
South	9	×	20.25	=	182.25
East	9	×	14.50	=	130.50
West	9	×	14.50	=	130.50
Total square feet in room				=	625.50

Step 2. Calculate the total square feet of areas not to be painted.

Area not to be painted	Height, in feet	×	Width, in feet	=	square feet in part
Doorway	7	×	3.50	=	24.50
Picture window	4.50	×	7	=	31.50
Window	3	×	4	=	12.00
Window	3	×	4	=	12.00
Total square feet not to be painted				=	80.00

Step 3. Calculate the square feet of wall space to be painted.

Square feet to be painted = total square feet in room − square feet not to be painted

Square feet to be painted = 625.50 − 80.00

Square feet to be painted = 545.50

(*b*) Calculate the number of gallons of paint needed.

Gallons needed = square feet to be painted ÷ square feet covered per gallon

Gallons needed = 545.50 ÷ 450

Gallons needed = 1.21

(*c*) Calculate the cost of the paint.

It appears that it would be wise for Julie to purchase 1 gallon and 1 quart (which makes 1.25 gallons) of paint.

Cost of paint = (No. of gallons × cost per gallon) + (No. of quarts × cost per quart)

Cost of paint = (1 × \$9.50) + (1 × \$2.75)

Cost of paint = \$9.50 + \$2.75

Cost of paint = \$12.25

This example assumes, of course, that the paint will actually cover the area stated on the container's label, which will depend largely upon how the painter applies the paint.

Calculating the Cost of Wallpapering

Wouldn't it be nice to replace that faded, out-of-date wallpaper you're so tired of? Perhaps one of the things holding you back is that you're afraid it will cost too much, particularly since the wallpaper you really like costs $30 a roll. Well, let's find out.

The first step in determining how much it will cost to wallpaper a wall or an entire room is to calculate the number of square feet of wall space to be covered. The process for doing so is to calculate the total square feet of the walls to be papered and then deduct the square feet of wall openings, such as doors and windows, that will not be covered. The process for doing that is explained in steps 1 to 9 in the preceding section on calculating the number of gallons needed to paint a surface. Refer to that step-by-step procedure if necessary.

After you have calculated the total square feet to be wallpapered, the next step is to determine how many rolls of wallpaper are needed. You do that by dividing the total square feet to be papered by the number of square feet in a roll of wallpaper. As a general rule, a single roll of wallpaper that does not have a recurring pattern that must be matched—as a stripe pattern or a solid color—will cover about 35 square feet. A double roll will cover approximately 70 square feet.

If the wallpaper has a recurring pattern that must be matched, a roll will not cover as many square feet as would one without a recurring pattern. That is because extra wallpaper may have to be cut off at the top or bottom to get the pattern to match when the paper is attached to the wall.

On the average, a pattern is repeated about every 18 inches. A roll of such wallpaper will cover about 30 square feet. If the pattern is repeated less often, say, every 30 inches, a roll will cover less, say, perhaps 27 or 28 square feet. As with no-match wallpaper, a double roll will cover twice the area a single roll will cover.

After the number of rolls of wallpaper needed is calculated, the cost for wallpapering the room can be determined by multiplying the cost per roll by the number of rolls required. If a wallpaper hanger is to be hired to do the job, that cost must also be included. Wallpaper hangers charge either an hourly rate or a flat charge per roll installed.

EXAMPLE Andrea Grems plans to wallpaper two walls in her living room. The north wall is 23 feet long, and the east wall is 14 feet long. Each wall is 9 feet high. There is one door, containing 25 square feet, in the east wall. The wallpaper Andrea has chosen has a striped pattern and a single roll will cover approximately 35 square feet. The roll will cost $20. A wallpaper hanger will be hired to do the job for a fee of $10 per roll installed. Calculate (*a*) the square feet of wall space to be papered, (*b*) the number of rolls of wallpaper needed, and (*c*) the cost of the wallpaper and installation.

Solution: (*a*) Calculate the square feet of wall space to be papered.

Step 1. Calculate the total square feet of wall space.

Wall	Height, in feet	×	length, in feet	=	square feet
North	9	×	23	=	207
East	9	×	14	=	126
	Total square feet in walls			=	333

Step 2. Calculate the number of square feet of wall space to be papered.

Total square feet to be papered = total square feet in walls − square feet not to be papered (door)

Total square feet to be papered = 333 − 25

Total square feet to be papered = 308

(*b*) Calculate the number of rolls of wallpaper needed.

Number of rolls needed	=	total square feet to be papered	÷	square feet covered per roll
Number of rolls needed	=	308	÷	35
Number of rolls needed	=		8.80	

(NOTE: Since partial rolls ordinarily cannot be bought, the number of rolls needed is rounded up to the next full roll. Therefore, 9 rolls of wallpaper are needed.)

(*c*) Calculate the cost of wallpaper and installation.

Step 1. Calculate the cost of the wallpaper.

Cost of wallpaper	=	number of rolls needed	×	cost per roll
Cost of wallpaper	=	9	×	$20
Cost of wallpaper	=		$180	

Step 2. Calculate the cost of installation.

Cost of installation	=	number of rolls needed	×	installation cost per roll
Cost of installation	=	9	×	$10
Cost of installation	=		$90	

Step 3. Calculate the complete cost of the wallpaper and installation.

Installed cost	=	cost of wallpaper	+	cost of installation
Installed cost	=	$180	+	$90
Installed cost	=		$270	

Since the cost of wallpaper varies widely, starting at prices below $10 per single roll, and many people do their own installation with the help of a friend, the cost of wallpapering a room can be considerably less than that shown in the preceding example.

Calculating Yards of Concrete

Concrete is measured and sold by the yard. In this instance, the term *yard* actually means *cubic yard,* which is an amount 3 feet long, 3 feet wide, and 3 feet high. Therefore, a cubic yard contains 27 cubic feet, as calculated by multiplying the length by the width by the height of the area (3 feet × 3 feet × 3 feet = 27 cubic feet). See Fig. 5-6.

To determine the number of cubic yards of concrete needed to cover an area and the cost of the concrete, proceed as follows:

1. Determine the length and width of the area to be covered, in feet.
2. Determine the thickness of the concrete to be used, in feet. For instance, a thickness of 3 inches is $\frac{1}{4}$ foot, a thickness of 4 inches is $\frac{1}{3}$ foot, and so on.
3. Determine the number of cubic feet of concrete to be used by multiplying the length of the area to be covered by the width of the area and by the thickness of the concrete, in feet.

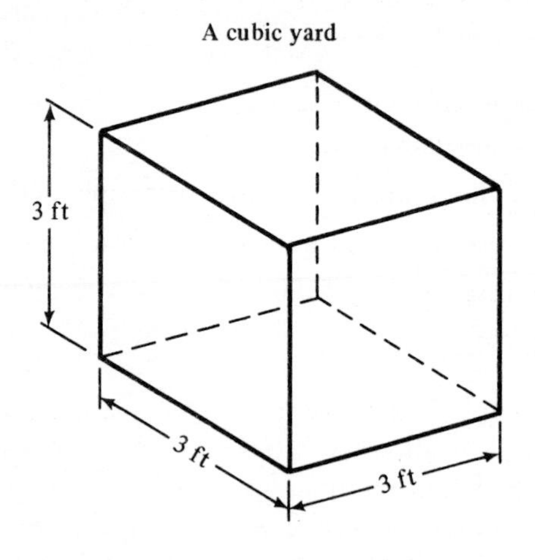

One cubic yard = 27 cubic feet

FIG. 5-6

4. Divide your answer from step 3, the number of cubic feet of concrete needed to cover your area, by 27, which is the number of cubic feet in a cubic yard. The result is the number of cubic yards of concrete needed.
5. Multiply your answer from step 4, the number of cubic yards of concrete needed, by the cost of concrete per cubic yard. The result is the cost of covering your area with concrete.

EXAMPLE Les Laidig plans to build a concrete patio measuring 10 feet by 15 feet, and he wants the concrete to be 3 inches thick. The cost of concrete is $45 per yard. Calculate (*a*) the cubic yards of concrete needed and (*b*) the cost for the concrete.

Solution: (*a*) Determine the number of cubic yards of concrete needed.

Step 1. Calculate the number of cubic feet in the area.

$$\text{Cubic feet in area} = \begin{matrix}\text{length}\\ \text{of area,}\\ \text{in feet}\end{matrix} \times \begin{matrix}\text{width}\\ \text{of area,}\\ \text{in feet}\end{matrix} \times \begin{matrix}\text{thickness}\\ \text{of concrete,}\\ \text{in feet}\end{matrix}$$

$$\text{Cubic feet in area} = 10 \times 15 \times \frac{1}{4}$$

$$\text{Cubic feet in area} = \frac{10}{1} \times \frac{15}{1} \times \frac{1}{4}$$

$$\text{Cubic feet in area} = \frac{150}{4}$$

$$\text{Cubic feet in area} = 37.50$$

Step 2. Calculate the number of cubic yards in the area.

$$\text{Cubic yards in area} = \begin{matrix}\text{cubic feet}\\ \text{in area}\end{matrix} \div \begin{matrix}\text{number of cubic feet}\\ \text{in a cubic yard}\end{matrix}$$

$$\text{Cubic yards in area} = 37.50 \div 27$$

$$\text{Cubic yards in area} = 1.39$$

(*b*) Calculate the cost of the concrete.

Cost of concrete = cubic yards in area × cost per cubic yard
Cost of concrete = 1.39 × $45
Cost of concrete = $62.55

Additional costs of concrete construction might also include sand or gravel for a base beneath the concrete and wire or reinforcing rods to be embedded in the concrete to add strength.

Calculating the Cost of Shingling a Roof

So you painted your house red and now it looks somewhat like a Christmas tree with those green shingles on the roof. Well, the shingles are old anyway, so why not replace them? But how much will it cost? That's easy to calculate, as you will see.

Shingles are ordinarily sold by the *square*. A square will cover an area 10 feet by 10 feet, or 100 square feet. To calculate how many squares will be needed, the first step is to calculate the number of square feet in the roof. Here are the steps:

1. If the house has a gable roof with identical sides, such as many ranch-style houses have, measure the length and the width of one side of the roof. Next, multiply the length by the width to calculate the number of square feet in one side of the roof. Then, multiply that answer by 2 to determine the number of square feet in the entire roof.
2. If the house has a roof with four equal sides that meet at a point in the middle, first calculate the number of square feet in one side of the roof. To do that, measure the distance from the peak to the outer edge of the roof. Then divide that distance in half and multiply it by the width of one side of the roof at its outer edge. The result is the number of square feet in one side of the roof. Multiply that amount by 4 to determine the total square feet in the entire roof.
3. If the roof has irregular shapes, calculate the number of square feet in each part of the roof and add the dimensions together to determine the roof's total square feet.

If you are unable to obtain actual dimensions by measuring the roof, you can make an estimate by measuring the roof's approximate length and width along the outside of the house at ground level. This will ordinarily be sufficient to provide you with a close approximation of the roof's area and cost of shingling. Of course, you may be able to obtain exact measurements from your home's blueprints or contractor's plans.

After you have calculated the total square feet in the roof, divide by 100 to determine the number of squares of shingles you will need. Often, the result will not be an even number of squares. The extra shingles needed beyond the even number of squares can be obtained by buying one or more *bundles* of shingles. A bundle is usually one-fourth or one-third of a square, depending upon the type of shingle and the shingle's dimensions.

Determine the cost per square of shingles by calling a local building supply center. Contractors ordinarily charge by the square for laying shingles, and you can call several of them to get an idea of labor costs.

Determine the total cost of shingling the roof by multiplying the number of squares needed by the cost per square of shingles and the cost of installation per square.

EXAMPLE Paul Simacek estimated each side of his ranch-style home with a gable roof to measure 50 feet by 16 feet. The cost of asphalt shingles is $40 per square. Schnidel Construction Company will install the shingles for $25 per square. Calculate the estimated cost of shingling the roof.

Solution:

Step 1. Calculate the number of square feet in one side of the roof.

Square feet in one side = length, in feet × width, in feet

Square feet in one side = 50 × 16

Square feet in one side = 800

Step 2. Calculate the number of square feet in the entire roof.

$$\text{Square feet in roof} = \begin{matrix}\text{square feet}\\ \text{in one side}\end{matrix} \times \begin{matrix}\text{number}\\ \text{of sides}\end{matrix}$$

$$\text{Square feet in roof} = 800 \times 2$$

$$\text{Square feet in roof} = 1{,}600$$

Step 3. Calculate the number of squares in the roof.

$$\text{Number of squares} = \begin{matrix}\text{square feet}\\ \text{in roof}\end{matrix} \div \begin{matrix}\text{square feet}\\ \text{in a square}\end{matrix}$$

$$\text{Number of squares} = 1{,}600 \div 100$$

$$\text{Number of squares} = 16$$

Step 4. Calculate the installed cost per square.

$$\text{Installed cost per square} = \begin{matrix}\text{cost of}\\ \text{shingles}\end{matrix} + \begin{matrix}\text{cost}\\ \text{of labor}\end{matrix}$$

$$\text{Installed cost per square} = \$40 + \$25$$

$$\text{Installed cost per square} = \$65$$

Step 5. Calculate the cost of shingling the roof.

$$\text{Total cost} = \begin{matrix}\text{installed cost}\\ \text{per square}\end{matrix} \times \begin{matrix}\text{number of}\\ \text{squares}\end{matrix}$$

$$\text{Total cost} = \$65 \times 16$$

$$\text{Total cost} = \$1{,}040$$

Shingling a roof with a gentle pitch is easy to do and you can save a considerable amount ($400 in this example) by doing the work yourself.

Calculating Board Feet

Lumber is sold by the *board foot.* A board foot is a piece of wood measuring 12 inches in length, 12 inches in width, and 1 inch in thickness. It contains 144 cubic inches, as calculated by multiplying the board's length by its width by its thickness ($12 \times 12 \times 1 = 144$). Lumber measurements reflect the lumber's dimensions in its rough state before planing and finishing. Therefore, a finished board is not quite as wide or thick as the stated measurement.

To determine the number of board feet in a piece of lumber, determine the number of cubic inches by multiplying the length by the width by the thickness, all in inches. Since the length of a board is usually stated in feet, convert that measurement to inches by multiplying it by 12 (12 inches per foot) before multiplying.

To calculate the price of a piece of lumber, multiply the number of board feet by the price per board foot.

EXAMPLE Larry Zimmerman plans to build a book case out of solid oak boards. He needs to buy two boards, each measuring 12 feet long, 12 inches wide, and 1 inch thick. The cost is $3.60 per board foot. Calculate (*a*) the number of board feet in the two boards and (*b*) the cost of the two boards.

Solution: (*a*) Calculate the board feet in the two boards.

Step 1. Convert a board's length to inches.

$$\text{Length, in inches} = \text{number of feet} \times \text{inches per foot}$$

$$\text{Length, in inches} = 12 \times 12$$

$$\text{Length, in inches} = 144$$

Step 2. Calculate the cubic inches in the two boards.

$$\text{Cubic inches in boards} = \left(\text{length, in inches} \times \text{width, in inches} \times \text{thickness, in inches}\right) \times \text{no. of boards}$$

$$\text{Cubic inches in boards} = (144 \times 12 \times 1) \times 2$$

$$\text{Cubic inches in boards} = 1{,}728 \times 2$$

$$\text{Cubic inches in boards} = 3{,}456$$

Step 3. Calculate the number of board feet.

$$\text{Board feet} = \text{cubic inches in boards} \div \text{cubic inches in a board foot}$$

$$\text{Board feet} = 3{,}456 \div 144$$

$$\text{Board feet} = 24$$

(*b*) Calculate the cost of the boards.

$$\text{Cost} = \text{number of board feet} \times \text{cost per board foot}$$

$$\text{Cost} = 24 \times \$3.60$$

$$\text{Cost} = \$86.40$$

Full-dimension, rough-sawn lumber can often be purchased directly from area sawmills at a price considerably lower than that paid at a building supply company for finished lumber. Usually, however, this route is economical only if you have your own planing and sanding equipment to turn the rough lumber into usable finished wood.

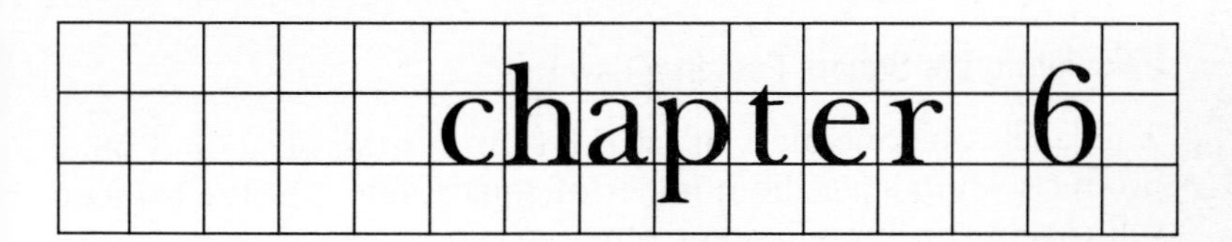

Real Estate Calculations

For most people, transactions involving real estate are among the highest in dollar amount and the most permanent. Often, the ability to make your own calculations is necessary to better understand your position or to verify the accuracy of information supplied to you by others.

Calculating the Square Feet in a Bare Residential Lot

Most residential lots are rectangular in shape. The number of square feet in the parcel is determined by multiplying the lot's length by its width, in feet.

EXAMPLE A residential lot is 80 feet wide and 132 feet deep. Calculate the number of square feet in it.

Solution:

$$\text{Square feet in lot} = \text{length, in feet} \times \text{width, in feet}$$
$$\text{Square feet in lot} = 132 \times 80$$
$$\text{Square feet in lot} = 10{,}560$$

This basic computation is often a part of the procedure followed in performing more complicated real estate calculations.

Calculating the Square Feet in a Lawn

The labels on containers of lawn fertilizer, grass seed, and other products often state the number of square feet that the contents will cover. If a house and/or other structures are situated on the lot, their area, in square feet, is deducted from the total square feet in the lot to determine the number of square feet of lawn space. The number of containers of fertilizer, seed, or whatever that are necessary to cover the lawn area can then be calculated as follows:

1. Calculate the number of square feet in the entire parcel of ground (as illustrated in the preceding example).
2. Calculate the number of square feet occupied by the house and/or structures as explained above.
3. From the total square feet in the parcel (as calculated in step 1), deduct the number of square feet occupied by various structures (as calculated in step 2). The result is the number of square feet in the lawn.
4. Divide the number of square feet in the lawn (as calculated in step 3) by the number of square feet the contents of the container will cover to determine how many containers of fertilizer, seed, or whatever to buy.

EXAMPLE Alvin Campbell plans to fertilize his lawn. The label on the bag of fertilizer states that the contents will cover 5,000 square feet. Campbell's lot measures 75 feet by 120 feet, and the house on the lot measures 50 feet by 30 feet. Determine (*a*) the number of square feet in the lawn and (*b*) the number of bags of fertilizer needed.

Solution: (*a*) Determine the square feet in the lawn.

Step 1. Calculate the total square feet in the lot.

Square feet in lot	=	length, in feet	×	width, in feet
Square feet in lot	=	120	×	75
Square feet in lot	=		9,000	

Step 2. Calculate the square feet occupied by the house.

Square feet in house = length, in feet × width, in feet

Square feet in house = 50 × 30

Square feet in house = 1,500

Step 3. Calculate the square feet in the lawn.

Square feet in lawn = square feet in lot − square feet in house

Square feet in lawn = 9,000 − 1,500

Square feet in lawn = 7,500

(*b*) Calculate the bags of fertilizer needed.

Number of bags needed = square feet in lawn ÷ square feet covered by a bag

Number of bags needed = 7,500 ÷ 5,000

Number of bags needed = 1.5 (or $1\frac{1}{2}$)

It can be seen that $1\frac{1}{2}$ bags, each capable of covering 5,000 square feet, are needed. Perhaps a bag covering 7,500 square feet can be located or two bags, one capable of covering 5,000 square feet and another of covering 2,500 square feet, can be purchased. Of course, with some products it is nice to have a little extra on hand, so purchasing a litttle more than actually needed may be wise.

CALCULATING THE ACRES IN A PARCEL OF LAND

Can you picture an acre of land? Technically, an acre contains 43,560 square feet, or it would be a square parcel measuring 208.71 feet on each side. These numbers may not help you develop an accurate image of an acre's size, however.

Perhaps the easiest way to visualize an acre is to think of a football field, which is 300 feet long, from goal line to goal line, and 160 feet wide. That is 48,000 square feet, or slightly more than an acre (about one-tenth more). In fact, if the area between the 10-yard line and the goal line is excluded, leaving a parcel of ground 270 feet (90 yards) by 160 feet the field measures one acre almost exactly. See Fig. 6-1.

To calculate the number of acres in a parcel of ground, proceed as follows:

1. Multiply the parcel's length by its width, both in feet, to determine the number of square feet in the area.
2. Divide the number of square feet in the parcel (as calculated in step 1) by the number of square feet in an acre, 43,560. The result is the acres in the parcel.

EXAMPLE A parcel of land measures 108 feet by 202 feet. How many acres is that?

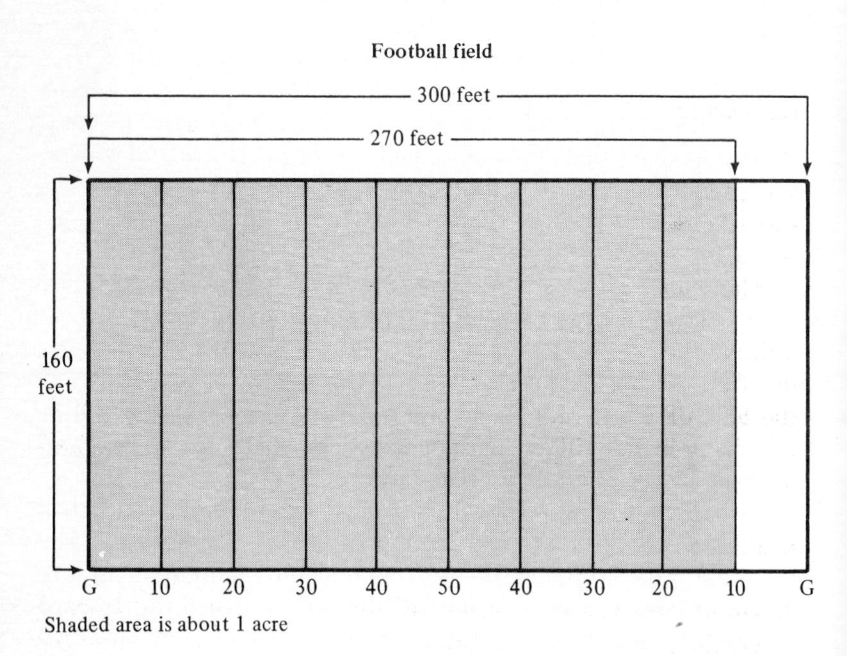

FIG. 6-1

Solution:

Step 1. Determine the square feet in the parcel.

$$\text{Square feet in parcel} = \text{parcel's length, in feet} \times \text{parcel's width, in feet}$$

$$\text{Square feet in parcel} = 202 \times 108$$

$$\text{Square feet in parcel} = 21{,}816$$

Step 2. Determine the number of acres.

$$\text{Acres} = \text{square feet in parcel} \div \text{square feet in an acre}$$

$$\text{Acres} = 21{,}816 \div 43{,}560$$

$$\text{Acres} = .5008 \text{ (about } \tfrac{1}{2}\text{) acre}$$

This procedure works well for calculating the number of acres in a relatively small parcel of land. For larger areas, another procedure is often used. It is described in the following section.

Calculating Acres in Rural Land

Farms, ranches, and other rural land in most parts of the country are measured in sections or portions of sections. A section is a square piece of land measuring 1 mile on each side and containing 640 acres. See Fig. 6-2.

The legal description of a parcel of rural land that is smaller than a section identifies the location of the property but does not indicate the number of acres in the parcel. For instance, the legal description for a 40-acre parcel would be as follows: The NW $\frac{1}{4}$ of the SE $\frac{1}{4}$ of Section 8. The number of acres in the parcel can be calculated by two different methods, both of which are described below.

Method 1

1. Draw a square to represent the section. Label the directions north, south, east, and west to avoid confusion in locating the subdivisions.

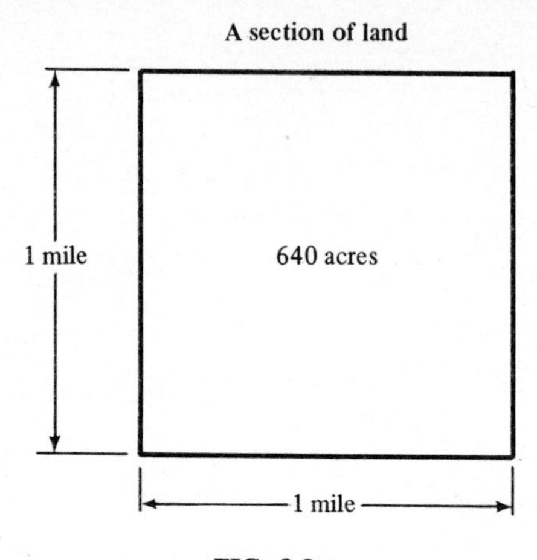

FIG. 6-2

2. Locate the first subdivision of the section by starting at the *end* of the legal description and reading toward the front. Sketch this subdivision on the square you drew in step 1. (*Note:* Legal descriptions are read from front to back, but a parcel of ground is located by starting at the end of the legal description and working toward the beginning.)
3. Continue to the next subdivision, again reading from the end of the legal description toward its beginning. Sketch this subdivision on the square representing the section.
4. Continue the process of identifying each subdivision and sketching it on your map of the section until you have identified the parcel of ground in the legal description.
5. As each subdivision is marked on your sketch, figure out how many acres are in the parcel, using the original 640 acres in the section as your beginning point. To do that, multiply the fraction as shown in the subdivision's legal description by the remaining acres in the parcel.

EXAMPLE Calculate the number of acres in a parcel of land described as follows: The S $\frac{1}{2}$ of the NE $\frac{1}{4}$ of the NW $\frac{1}{4}$ of Section 10.

Solution: Refer to Fig. 6-3 on page 94 as you complete each step.

Step 1. Draw a square, which will represent Section 10. Label the directions.

Step 2. Starting at the back of the legal description, identify the first subdivision (NW $\frac{1}{4}$). Sketch the subdivision on the map of Section 10. Identify the number of acres in this subdivision.

Step 3. Continue toward the beginning of the legal description and identify the next subdivision (NE $\frac{1}{4}$). Sketch the subdivision on the map and identify the number of acres in the subdivision.

Step 4. Continue toward the beginning of the legal description and identify the next subdivision (S $\frac{1}{2}$). Sketch the subdivision on the map and identify the number of acres in the subdivision.

Method 2

1. Multiply the fractions of all of the subdivisions by each other.
2. Multiply the answer from step 1 by the number of acres in a section, 640.

EXAMPLE Calculate the number of acres in a parcel of land described as follows: The S $\frac{1}{2}$ of the NE $\frac{1}{4}$ of the NW $\frac{1}{4}$ of Section 10. (*Note:* This is the same legal description as used in the preceding example.)

Solution:

Step 1. Multiply the fractions of all of the subdivisions.

$$\text{Legal description: S } \tfrac{1}{2} \text{ of the NE } \tfrac{1}{4} \text{ of the NW } \tfrac{1}{4}$$

$$\text{Fractions: } \frac{1}{2} \times \frac{1}{4} \times \frac{1}{4} = \frac{1}{32}$$

Step 2. Multiply the fractional equivalent of the subdivisions by the number of acres in a section.

$$\text{Acres in parcel} = \begin{array}{c}\text{fractional equivalent}\\ \text{of subdivisions}\end{array} \times \begin{array}{c}\text{acres in}\\ \text{a section}\end{array}$$

$$\text{Acres in parcel} = \frac{1}{32} \times 640$$

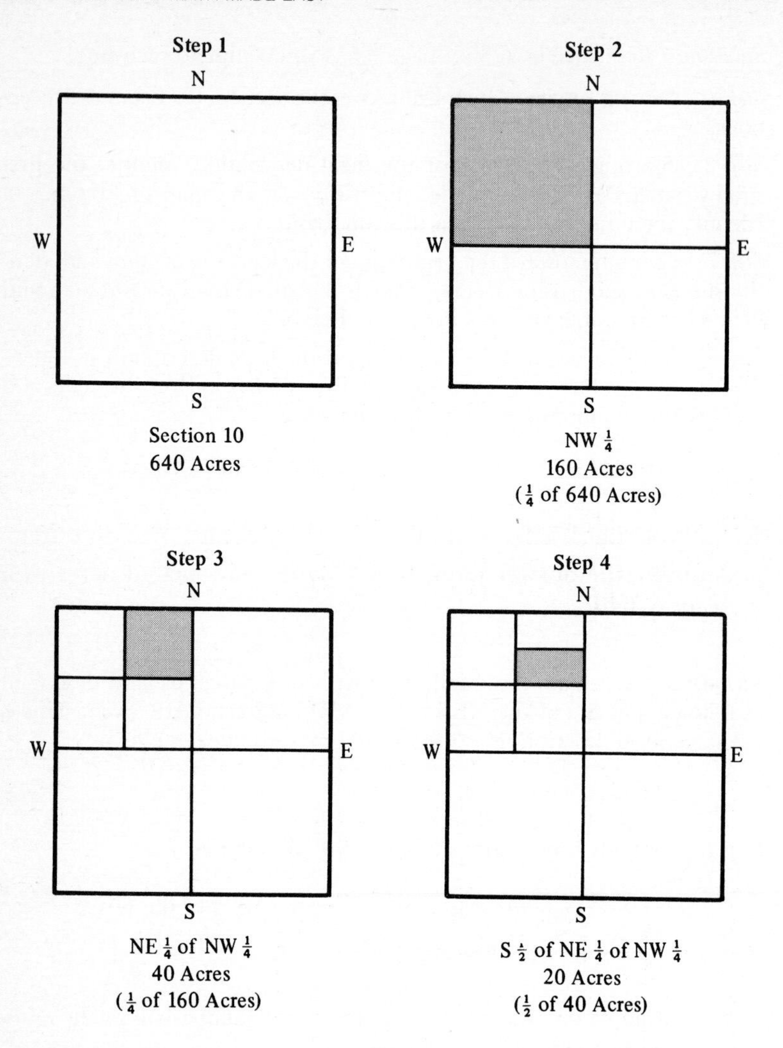
Step 1
N
W
E
S
Section 10
640 Acres
Step 2
N
W
E
S
NW $\frac{1}{4}$
160 Acres
($\frac{1}{4}$ of 640 Acres)
Step 3
N
W
E
S
NE $\frac{1}{4}$ of NW $\frac{1}{4}$
40 Acres
($\frac{1}{4}$ of 160 Acres)
Step 4
N
W
E
S
S $\frac{1}{2}$ of NE $\frac{1}{4}$ of NW $\frac{1}{4}$
20 Acres
($\frac{1}{2}$ of 40 Acres)

FIG. 6-3

$$\text{Acres in parcel} = \frac{1}{32} \times \frac{640}{1}$$

$$\text{Acres in parcel} = \frac{640}{32}$$

$$\text{Acres in parcel} = 20$$

In reading the legal description of rural land, be watchful for the word "and," which indicates there are two separate parcels of ground. For example, "The NW $\frac{1}{4}$ and the SE $\frac{1}{4}$ of the SW $\frac{1}{4}$ of Section 15," identifies two parcels of land. The first, the NW $\frac{1}{4}$, is 160 acres. The second, the SE $\frac{1}{4}$ of the SW $\frac{1}{4}$, is 40 acres. Therefore, a total of 200 acres is identified in the legal description.

Townships

In each of the preceding legal descriptions, every section was identified by a number. A section's number refers to its location within a township. A township contains 36 square miles (36 sections), or 23,040 acres. Sections within a township are always numbered as shown in Fig. 6-4.

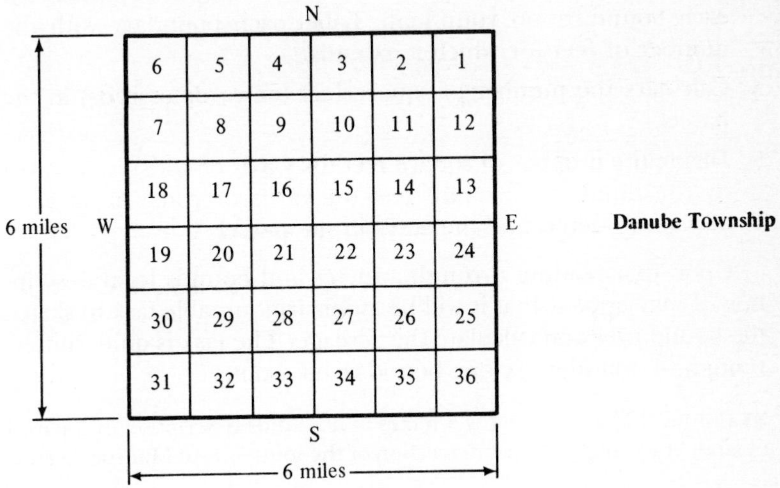

N

6	5	4	3	2	1
7	8	9	10	11	12
18	17	16	15	14	13
19	20	21	22	23	24
30	29	28	27	26	25
31	32	33	34	35	36

6 miles W — E **Danube Township**

S

6 miles

FIG. 6-4

Calculating Metes and Bounds Acres

Rural land in most of the eastern United States and Texas (and occasionally in other states too) is often identified through a system known as *metes and bounds*. This type of legal description starts at a landmark such as a large rock or a man-made identification marker known as the *point of beginning* (POB). From there the legal description identifies the boundaries of the parcel of land by directions and measurements. The boundary of a metes and bounds legal description always returns to the point of beginning to enclose an area.

The number of acres contained in a parcel of land identified by a metes and bounds description can be calculated as follows:

1. Reading from the *front* of the metes and bounds description, identify the point of beginning. Draw a map and identify the POB.
2. Read the first portion of the legal description. Sketch that boundary on your map and label it with the number of feet for which that boundary extends.
3. Continue to read each portion of the description, and sketch each boundary on your map. Label each boundary with the number of feet for which it extends.
4. Calculate the number of square feet (or yards or rods) in the parcel.
5. Divide the number of square feet (or yards or rods) in the area by the number of square feet (or yards or rods) in an acre, 43,560, to determine the acres in the parcel.

Upon first reading through a metes and bounds legal description, it may appear that it will be an insurmountable task to sketch the boundaries and calculate the acreage. The task is quite simple, though, if you identify one boundary at a time.

EXAMPLE The following is a metes and bounds description of a parcel of land: "Beginning at the intersection of the south line of Monroe Avenue and the east line of State Street, thence south along the east line of State Street 90 feet; thence east 121 feet on a course parallel to the south line of Monroe Avenue; thence north 90 feet on a course parallel to the east line of State Street to the south line of Monroe Avenue; thence west 121 feet along the south line of Monroe Avenue to the point of beginning." (*a*) Draw a sketch of the area identified by the metes and bounds description and label each boundary's measurements. (*b*) Calculate the number of acres in the parcel.

Solution: (*a*) Draw a sketch of the parcel and label the boundaries. As you take each step suggested below, refer to Fig. 6-5.

Step 1. Draw a sketch and label it.

Step 2. Mark the point of beginning with an X. The POB is identified in the description with the words, "Beginning at the intersection of the south line of Monroe Avenue and the east line of State Street."

Step 3. Sketch the first boundary of the description. This boundary is identified as, "thence south along the east line of State Street 90 feet."

Step 4. Sketch the next boundary of the description. This boundary is identified as, "thence east 121 feet on a course parallel to the south line of Monroe Avenue."

Step 5. Sketch the next boundary of the description. This boundary is identified as, "thence north 90 feet on a course parallel to the east line of State Street to the south line of Monroe Avenue."

Step 6. Sketch the next boundary of the description. This boundary is identified as, "thence west 121 feet along the south line of Monroe Avenue to the point of beginning."

(*b*) Calculate the acres in the parcel.

Step 1. Calculate the square feet in the parcel.

Square feet in parcel = length, in feet × width, in feet

Square feet in parcel = 121 × 90

Square feet in parcel = 10,890

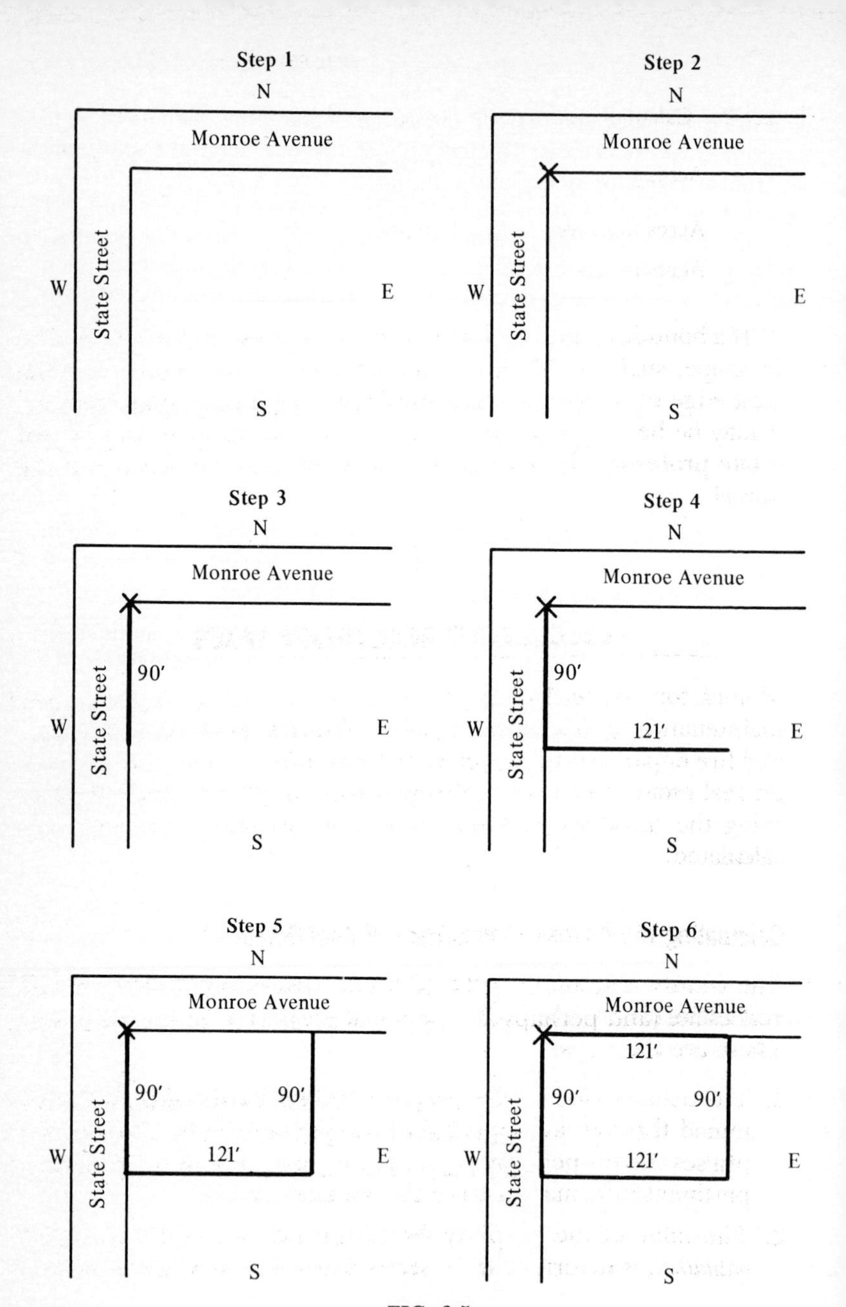
Step 1
N
Monroe Avenue
State Street
W
E
S
Step 2
N
Monroe Avenue
State Street
W
E
S
Step 3
N
Monroe Avenue
90′
State Street
W
E
S
Step 4
N
Monroe Avenue
90′
121′
State Street
W
E
S
Step 5
N
Monroe Avenue
90′
90′
121′
State Street
W
E
S
Step 6
N
Monroe Avenue
121′
90′
90′
121′
State Street
W
E
S

FIG. 6-5

Step 2. Calculate the acres in the parcel.

Acres in parcel	=	square feet in parcel	÷	square feet in an acre
Acres in parcel	=	10,890	÷	43,560
Acres in parcel	=	.25 (or $\frac{1}{4}$) acre		

If a boundary line of a metes and bounds description is irregular in shape, such as, "thence in a southeasterly direction along the west edge of Maverick Creek until reaching a large granite rock," it may be best to seek the services of your county assessor, a real estate professional, or a surveyor to determine the acreage of the parcel.

CALCULATING REAL ESTATE TAXES

Monies for city and county services such as street repairs, road maintenance, public schools, public libraries, police departments, and fire departments are generated through the collection of taxes on real estate. In some locales, personal property also is taxed by using the same tax rate and procedures as real estate taxes are calculated.

Calculating the Assessed Valuation of Real Estate

The county assessor is responsible for determining the value of real estate (and perhaps also personal property) for tax purposes. These are the steps:

1. The actual value of the property, called *market value,* is determined through an appraisal of the property. The assessor appraises the property by personally inspecting it or by gathering pertinent information from the property owner.
2. The value of the property for tax purposes, called the *assessed valuation,* is determined. In some states, assessed valuation is at

100% of market value; in others, it ranges from 25% to 60% of market value. Contact your county assessor to determine the rate at which property is assessed in your area.

To calculate the assessed valuation of your real estate, multiply the market value, as determined by the assessor, by the assessment rate.

EXAMPLE In the state in which Mariam Holland lives, real estate is assessed at 60% of market value. The county assessor appraised Mariam's house and determined its market value to be $100,000. Determine the property's assessed valuation.

Solution:

Assessed valuation	=	market value	×	assessment rate
Assessed valuation	=	$100,000	×	60%
Assessed valuation	=	$100,000	×	.60
Assessed valuation	=		$60,000	

It should be noted that a property's assessed valuation, as determined by the county assessor, is often *not* a good indicator of the amount that the property would sell for if it were put up for sale. That is because the assessment rate may be less than 100% of the market value, as indicated above, and because the assessor's most recent inspection of the property may have been many years ago—long before recent increases or decreases in the property's value.

Calculating the Amount of Property Tax

The rate at which property is taxed, called the *tax rate,* is stated in dollars per $1,000 of assessed valuation, dollars per $100 of assessed valuation, cents per dollar of assessed valuation, or mills per dollar of assessed valuation. A *mill* is one-tenth of one cent, or one-thousandth of a dollar, and as a decimal it is expressed as .001. To calculate the amount of real estate tax on a piece of property, proceed as follows:

1. Determine the property's assessed valuation and the tax rate by contacting the county assessor.

2. Apply the tax rate to the property's assessed valuation to determine the amount of real estate tax, as follows:

   a. If the tax rate is stated in cents per dollar of assessed valuation, state the tax rate in dollars rather than cents. For example, 7 cents becomes $.07, 8.5 cents becomes $.085, and so on. (This is done by removing the cents sign, moving the decimal point in the amount two places to the left, and adding the dollar sign.) Determine the real estate tax by multiplying the tax rate per dollar by the property's assessed valuation.

   b. If the tax rate is stated in dollars per $100 of assessed valuation, divide the property's assessed valuation by 100 before multiplying by the tax rate.

   c. If the tax rate is stated in dollars per $1,000 of assessed valuation, divide the property's assessed valuation by 1,000 before multiplying by the tax rate.

   d. If the tax rate is stated in mills per dollar, divide the number of mills by 1,000 (by moving the decimal point three places to the left) before multiplying by the property's assessed valuation. For example, 54 mills becomes .054; 60 mills becomes .060; and so on.

EXAMPLE 1 The assessed valuation of a house is $50,000, and the tax rate is 7 cents per dollar. Calculate the amount of real estate tax on the property.

Solution:

Property tax = assessed valuation × tax rate per $1

Property tax = $50,000 × 7¢

Property tax = $50,000 × $.07

Property tax = $3,500

EXAMPLE 2 The assessed valuation of a house is \$60,000, and the tax rate is \$9.25 per \$100 of assessed valuation. Calculate the amount of real estate tax on the property.

Solution:

$$\text{Property tax} = (\text{assessed valuation} \div 100) \times \text{tax rate per } \$100$$
$$\text{Property tax} = (\$60{,}000 \div 100) \times \$9.25$$
$$\text{Property tax} = \$600 \times \$9.25$$
$$\text{Property tax} = \$5{,}550$$

EXAMPLE 3 School officials state that the construction of a new elementary school will raise annual real estate taxes "only" \$1.20 per \$1,000 of assessed valuation. Calculate the annual increase in real estate taxes on a home assessed at \$80,000.

Solution:

$$\begin{array}{c}\text{Property tax}\\\text{increase}\end{array} = (\text{assessed valuation} \div 1{,}000) \times \begin{array}{c}\text{tax rate}\\\text{per } \$1{,}000\end{array}$$
$$\begin{array}{c}\text{Property tax}\\\text{increase}\end{array} = (\$80{,}000 \div 1{,}000) \times \$1.20$$
$$\begin{array}{c}\text{Property tax}\\\text{increase}\end{array} = \$80 \times \$1.20$$
$$\begin{array}{c}\text{Property tax}\\\text{increase}\end{array} = \$96$$

EXAMPLE 4 The assessed valuation of a house is \$120,000, and the tax rate is 75 mills per dollar of assessed valuation. Calculate the real estate tax on the property.

Solution:

$$\text{Property tax} = \text{assessed valuation} \times (\text{tax rate per } \$1 \div 1{,}000)$$
$$\text{Property tax} = \$120{,}000 \times (75 \text{ mills})$$
$$\text{Property tax} = \$120{,}000 \times (75 \div 1{,}000)$$

Property tax = $120,000 × $.075

Property tax = $9,000

CALCULATING THE CASH NEEDED TO PURCHASE A HOME

It's still the American dream—owning a home. But can you afford it? This section will help you answer that question by providing a step-by-step process for identifying all of your potential cash expenditures, and also some possible sources of cash receipts, that are normal when buying a home.

Potential Cash Expenditures

The costs involved in buying a home will vary a great deal depending upon whether cash is paid or a loan is obtained. Furthermore, other costs will vary from one transaction to another. For those reasons, it is impossible to indicate exact amounts here, but areas in which costs might be incurred are as follows:

1. ***Down payment.*** If you will be obtaining a loan to finance the purchase of a home, you will ordinarily be required to make a down payment of 5% to 25% of the purchase price. Thus, if you are buying a $100,000 home and the lender requires a 20% down payment, you will need $20,000 cash ($100,000 × 20% = $20,000); the lender will loan you the remaining $80,000.
2. ***Property appraisal fee.*** If a loan is obtained to finance the purchase, the lender will ordinarily charge the buyer a fee for appraising the property to determine whether it is worth at least the amount being loaned on it. Depending upon the type of property and local trends, this might amount to $100 to $300 or more.
3. ***Loan origination fee.*** Some lenders charge a *loan origination fee* of perhaps 1% to 3% of the amount borrowed. In essence, this amounts to a service charge for processing your loan. Thus, if

you obtain a $50,000 loan and a 2% loan origination fee is charged, the fee will be $1,000 ($50,000 × 2% = $1,000).

4. ***Discount points.*** Some lenders charge the buyer *discount* points when a loan is obtained. This might be 1% or more of the amount borrowed. Often, discounts are referred to simply as *points*. Each "point" is 1%. Thus, if you obtain a $60,000 loan and "one point" is assessed, you must pay $600 ($60,000 × 1% = $600).
5. ***Title opinion.*** Before completing your purchase, you will need to determine if the seller in fact owns the property and if there are any liens, judgments, or other assessments against the property that would limit your ownership in the property. Ordinarily, the seller updates the property's abstract (record of the property's history) or other evidence of title. Then the buyer hires an attorney to review the abstract and render an opinion as to the condition of the title. Depending upon the amount of work involved by an attorney, this may cost from $50 to $150 or more.
6. ***Recording fees.*** After the purchase is completed, your title to the property (the deed) should be recorded in the county courthouse. The recording fee is usually minimal, perhaps $10 to $25.
7. ***Property insurance premium.*** If you obtain a loan when purchasing the property, the lender will undoubtedly require you to obtain insurance on the property. Even if you paid cash for the property, you would most likely obtain insurance anyway. Ordinarily, a full year's or a half year's property insurance premium is payable in advance. The premium will vary with several factors including the value of your property, your location, and the type of coverage obtained. As a rough guideline, the premium for a homeowner's policy on a $100,000 home might be in the range of $375 to $450 per year.
8. ***Utility deposits.*** In most communities, you must make a deposit with the utility companies providing electrical, natural gas, water, and sewer services before those services will be hooked up. The deposit varies, but often an amount equal to the highest 2 months'

costs from the preceding year is required. Therefore, utility deposits may well amount to several hundred dollars.

However, if you have already established a reliable record for paying utility bills in the same community, most likely no deposit will be required.

9. ***Other items.*** There may be other amounts which the buyer of real estate will need to pay at the time the transaction is closed. Most of them will be identified in your contract with the seller. They may include such items as paying the seller for fuel oil or liquid propane (LP) gas in a tank or paying the seller for various pieces of personal property not included in the purchase price of the home.

Since the costs incurred when purchasing a home can vary widely with your circumstances, it is wise to check with the lender, insurance company, utility companies, and others before making an offer to buy.

Potential Cash Receipts

Under certain circumstances, it is possible that the buyer of real estate may receive cash from the seller at the time the purchase is completed, as illustrated below:

1. ***Prorated rent receipts.*** If you are buying real estate which includes rental property (say, half of a duplex) on which the seller has collected rents in advance, it is normal for the buyer to receive a pro rata amount of the prepaid rents from the seller. Assume, for example, that, on the first day of a 30-day month, the seller collected $300 rent from a tenant as rent paid in advance for that month. Assume further that you complete the purchase of the property on the eleventh of the month. Therefore, the seller owned the property for 10 days, or one-third of the month, and is entitled to one-third of the rent, or $100. Since you will own the property for 20 days, or two-thirds

of the month, you are entitled to two-thirds of the rent, or $200. The seller should pay you that amount when the sale is closed.

2. ***Security deposits.*** If you are buying real estate which includes rental property on which the seller has collected security deposits from the tenants, the seller can transfer the deposits to you. Otherwise, the seller can refund the deposits to the tenants and you can then collect your own security deposit from them. In either case, even though you receive the money, it is usually not yours to use. Many states require that security deposits be placed in a separate bank account designated for that purpose.
3. ***Other.*** Under rare circumstances, there may also be other cash amounts that the buyer will receive upon completion of the purchase.

Summary of Cash Needed

After gathering all of your information, you can complete a form like that shown in Table 6-1 to identify the amount of cash you will need to finalize your purchase.

CALCULATING NET PROCEEDS FROM THE SALE OF REAL ESTATE

Let's assume the following: (1) You have just sold your home for $100,000. (2) You have a $40,000 loan on the home which must be paid off. (3) Baker Real Estate Agency handled the sale for you and charges you a 7% commission, which amounts to $7,000.

So will you end up with $53,000 ($100,000 less $40,000 and $7,000 equals $53,000) when the sale is completed? One might asume that the net proceeds would be near that amount, but many other costs may be involved. There may be some other cash receipts that must be considered. The amounts can vary widely with the circumstances, but the following lists contain many of the possibilities.

TABLE 6-1 CASH NEEDED TO PURCHASE A HOME

Potential cash payments		
1. Down payment	$ ______	
2. Property appraisal fee	______	
3. Loan origination fee	______	
4. Discount points	______	
5. Title opinion	______	
6. Recording fees	______	
7. Property insurance premium	______	
8. Utilities deposits	______	
9. Other: ______	______	
Other: ______	______	
Other: ______	______	
A. Total cash expenditures		$ ______
Potential cash receipts		
1. Prorated rent receipts	$ ______	
2. Security deposits	______	
3. Other: ______	______	
Other: ______	______	
B. Total cash receipts		$ ______
Total cash needed (A minus B)		$ ______

Potential Cash Receipts

Upon the sale of your home, cash might be received from several sources including the following:

1. ***Sales price.*** The sales price is, of course, the major source of cash you will receive from the transaction. From it, and any other cash you receive, will be deducted the various cash expenditures connected with the sale.

2. ***Escrow account refund.*** If you currently have a loan on the property, each monthly payment made to the lending institution might include one-twelfth of your annual property insurance premium and one-twelfth of your annual real estate property taxes. The lender accumulates these amounts in a separate account, called an *escrow account,* and then pays the insurance premium and real estate taxes for you when they become due. Therefore, at the time your sale is closed, there may be amounts accumulated in your escrow account, which should be refunded to you. The lender should be able to tell you the exact amount in the escrow account.

3. ***Property insurance premium rebate.*** Since annual or semiannual property insurance premiums are ordinarily paid in advance, there may be prepaid insurance premiums at the time your sale is completed. You can therefore cancel the insurance policy and receive a refund for the unused time period for which you prepaid.

   Ordinarily, when a property insurance policy is canceled, the insurance company will make a refund on a less than prorata basis. For instance, assume you paid a full year's insurance premium of $400 in advance. Now, exactly 6 months after making that payment, you sell your home and cancel the policy. Instead of refunding exactly half of the premium paid, $200, the insurance company will probably refund a lesser amount, say, $175.

   Therefore, it is usually wise to attempt to have the buyer assume your existing policy, with the insurance company's permission, and pay you the exact amount of remaining prepaid premium. In our example, therefore, the buyer would pay you $200 for the remaining half year's prepaid insurance premium.

4. ***Sale of personal property.*** Often, the sale of real estate does not include certain personal property located on the premises, such as fuel oil or LP gas in the tank, unless your contract with the buyer specifically states that it is included. Therefore, since the buyer will need those items anyway, it is normal procedure for the buyer to purchase them from the seller, separately from the real estate transaction.

5. ***Utility deposit refunds.*** If you have made utility deposits for electrical, natural gas, water, sewer, and other services, those amounts should be refunded to you by the utility companies. If you are staying in the same community, however, the deposits most likely will be transferred to your new location.
6. ***Other.*** It is possible that there might also be other sources of cash receipts, depending upon your individual circumstances.

Potential Cash Expenditures

Following are items for which you may incur cash expenditures upon the sale of your home.

1. ***Loan balance repayment.*** If you currently have a loan on the property, that loan ordinarily must be paid off by you at the time of the sale. You can check with your lender to determine the exact balance that will remain on the date the sale is completed.
2. ***Loan prepayment penalty.*** Many loan agreements include a *prepayment penalty clause.* This clause gives the lender the right to charge you a penalty of, say, 1% or 2% of your current loan balance if you pay off the loan in advance of its due date. Thus, if your loan balance is $40,000 and a 1% prepayment penalty is assessed, you will need to pay the lender $400 ($40,000 × 1% = $400). Check your loan agreement to determine if it includes such a stipulation. If it does, ask your lender if it will be enforced. Often it is not.
3. ***Real estate commission.*** If a real estate agency handles the sale for you, a sales commission will be charged. Ordinarily it is a percent of the sales price that is identified in your listing agreement with the agency.
4. ***Prorated real estate taxes.*** Real estate taxes are usually paid in arrears; that is, they are paid after the time period to which they apply. Assume, for example, that your real estate taxes are $1,200 per year (an average of $100 per month) and that you can pay them in two installments of $600 each on March 31 and September 30. If you make your scheduled $600 pay-

ment promptly on September 30, 1984, that payment is, in essence, the taxes on the property for a six-month period from July 1, 1983 to December 31, 1983. Therefore, even though you have just made the most recent tax payment required of you, taxes are still unpaid on the property for the period from January 1, 1984 through September 30, 1984.

Thus, if you were to sell the property on October 1, 1984, there would be $900 of real estate taxes in arrears for the period from January 1, 1984 to September 30, 1984 (9 months at $100 per month). It is ordinarily the seller's responsibility to bring the real estate taxes up to date as of the day on which the sale is completed; therefore, a $900 payment would be required of the seller.

It may be best to ask your county treasurer or a real estate professional to assist you in calculating the amount of prorated real estate taxes that you will need to pay on the projected sales date.

5. ***Special assessments.*** If taxes, called *special assessments,* have been levied on the property for the paving of a street, sewer construction, or other purpose, the seller may have to pay part or all of the past and/or future amounts due. Your sales contract with the buyer will indicate your responsibility here. You can check with the county treasurer or city clerk to determine the amount of any special assessments levied against your property.
6. ***Unpaid liens and judgments.*** If a lien or judgment has been filed against your property, say, by a contractor or repairman, it must ordinarily be released before the buyer will accept the property. That means you will have to pay whatever amount is stated in the lien or judgment or in some other way negotiate a release. If a lien or judgment has been filed against your property, you are probably aware of it; otherwise, you can check the county recorder's records.
7. ***Abstract continuation.*** The abstract or a similar document contains the history of title to your property. The title evidence must be brought up to date, ordinarily at the seller's expense, before the buyer will accept the property. The cost will vary

with when the abstract was last updated and how much work is required of the abstractor to search out and record any changes since that time. Under normal circumstances, a cost of $75 to $150 might be expected.

8. ***Prorated rental income.*** Assume that you are selling real estate which includes rental property (say, half of a duplex) on which you have collected 1 month's rent of $300 in advance at the beginning of a 30-day month. Further assume that the property changes hands on the eleventh of that month. Therefore, you will have owned the property for 10 days that month (one-third of the month) and the buyer will own it for 20 days that month (two-thirds of the month). You are therefore entitled to one-third of the rent collected in advance, or $100, and the buyer is entitled to two-thirds of the rent, or $200. Thus, under normal circumstances, you must pay the buyer $200 of that $300 rent you collected in advance.

9. ***Loan discount points.*** If the buyer is obtaining a Federal Housing Administration (FHA) insured loan or Veteran's Administration (VA or GI) guaranteed loan to finance the purchase of the property, loan *discount points* are ordinarily assessed. Each discount point, or simply "point," amounts to 1 percent of the loan balance obtained. By law, the buyer cannot pay the discount points, so the seller must therefore pay them in order for the buyer to obtain the loan and complete the purchase. Four, five, six, or more points are not uncommon.

   If the buyer obtains an FHA or VA loan for $30,000 and 6 points are assessed to the seller, you will need to pay the lender $1,800 in cash ($30,000 × 6% = $1,800). It is extremely important, therefore, that you thoroughly investigate your potential costs and understand your circumstances before you accept an offer which states that the buyer is to obtain an FHA-insured or VA-guaranteed loan. Any savings and loan association or bank can provide you with a full explanation.

10. ***Lender-required repairs.*** If the buyer is obtaining a Federal Housing Administration (FHA) insured loan, Veteran's Administration (VA or GI) guaranteed loan, or Farmer's Home

Administration (FmHA) loan, the government agency involved will send someone to appraise your real property. In order for one of those government agencies to approve the loan, the property must meet certain structural requirements. If your property does not meet the agency's requirements, as many properties do not, you will probably be required to make various improvements to the property before the loan will be approved. On even a very livable and seemingly sound house, the cost of improvements may run into hundreds or even thousands of dollars. Therefore, it is very important that you do not agree to make the improvements before you know their nature and probable cost.

11. ***Attorney's fees.*** It may be necessary to seek an attorney's advice and assistance, particularly if you are handling the sale yourself rather than utilizing the services of a real estate agent. These fees might range from nothing to several hundred dollars.

12. ***Deed preparation.*** Since the deed is an extremely important document, it should be prepared by a professional such as an attorney. The deed ordinarily can be prepared quickly and the charge for this is low, perhaps in the $10 to $15 range.

    If you are selling the property on a *contract for deed,* more time is usually required to prepare the document. A charge of $50 to $100 or more might be expected.

13. ***Security deposit refunds.*** If you are selling real estate which involves rental property on which you have collected security deposits, those deposits should be refunded to the tenants or transferred to the buyer. Most likely, this will not actually constitute a cash expenditure for you, since most states require that security deposits be held in a separate bank account. Therefore, this will merely amount to a simple transfer of existing funds.

14. ***Transfer tax.*** All states require that a tax, called *transfer tax, revenue stamps,* or *documentary stamps,* be paid when title to real estate is transferred. This cost is usually borne by the seller. Although the amount varies from state to state, it is usually in

TABLE 6-2 NET PROCEEDS FROM THE SALE OF REAL ESTATE

Potential cash receipts

1. Sales price	$ ________	
2. Escrow account refund	________	
3. Property insurance premium rebate	________	
4. Sale of personal property	________	
5. Utility deposit refunds	________	
6. Other: ________	________	
Other: ________	________	
A. Total cash receipts		$ ________

Potential cash receipts

1. Loan balance repayment	$ ________	
2. Loan repayment penalty	________	
3. Real estate commission	________	
4. Prorated real estate taxes	________	
5. Special assessments	________	
6. Unpaid liens and judgments	________	
7. Abstract continuation	________	
8. Prorated rental income	________	
9. Loan discount points	________	
10. Lender-required repairs	________	
11. Attorney's fees	________	
12. Deed preparation	________	
13. Security deposit refunds	________	
14. Transfer tax	________	
15. Other: ________	________	
Other: ________	________	
B. Total potential cash payments		$ ________
Expected net proceeds from sale (A − B)		$ ________

the range of \$1.10 for each \$1,000 of the sales price. Therefore, if the sales price is \$150,000, your transfer tax cost could be \$165 (\$150,000 ÷ 1,000 × \$1.10 = \$165).

15. ***Other.*** It is possible that other cash payments may be required by your unique circumstances.

Summary of Net Proceeds

After you have gathered all of the information identified in this chapter, you should be able to complete Table 6-2 and make an accurate calculation of the net proceeds you will receive from the sale of your property.

Since many potential costs are involved in selling a parcel of real estate, you should thoroughly investigate your situation and gather information to determine where you stand financially before you sign a contract to sell your property.

chapter 7

Calculating Earnings

It is often wise to check an employer's calculations of your earnings to see that you are being paid all that you are entitled to receive. Similar calculations are wise when you are estimating the amount of weekly, monthly, or annual earnings that would result from a prospective job change or when you are analyzing whether working overtime or moonlighting (taking a second job) is worth the effort.

METHODS OF CALCULATING EARNINGS

Many methods are used to calculate earnings, and a particular method is common for a certain type of employment. *Gross earnings,* as calculated here, refers to total earnings before deductions for taxes and other purposes.

Wages

The term "wages" refers to earnings calculated at an hourly rate. Usually, the regular pay rate is earned for a minimum number of hours worked per day, often 8. *Overtime pay,* usually at $1\frac{1}{2}$ times the regular pay rate, is earned for hours worked in excess of the established minimum. A special pay rate, often at twice the regular rate, is paid by some employers when employees work on Sundays or holidays. To calculate gross earnings for a pay period, proceed as follows:

1. Multiply the number of hours worked at the regular pay rate by the rate per hour.
2. If hours were worked at an overtime rate, calculate the overtime rate per hour by multiplying the regular pay rate by the overtime factor (say, $1\frac{1}{2}$). Then calculate the overtime pay by multiplying the overtime pay rate by the number of overtime hours worked.
3. Add the regular earnings and overtime earnings together.

To make an estimate of annual earnings, multiply the earnings for a typical week by the number of weeks for which you will be paid during the year.

EXAMPLE Glenn Chastine earns $12 per hour and is paid time and a half for hours worked in excess of 40 in a week. Last week, Glenn worked 42 hours. Calculate Glenn's (*a*) gross earnings for the week and (*b*) the estimated earnings for a year in which he will be paid for 50 weeks' work, assuming his earnings per week will average that determined in part (*a*) of this example.

Solution: (*a*) Calculate the week's gross earnings.

Step 1. Calculate the regular earnings.

Regular earnings = regular rate × regular hours
Regular earnings = $12.00 × 40
Regular earnings = $480.00

Step 2. Calculate the overtime rate.

Overtime rate = regular rate × overtime factor
Overtime rate = $12.00 × $1\frac{1}{2}$
Overtime rate = $12.00 × 1.5
Overtime rate = $18.00

Step 3. Calculate the overtime earnings.

Overtime earnings = overtime rate × overtime hours
Overtime earnings = $18.00 × 2
Overtime earnings = $36.00

Step 4. Calculate the week's gross earnings.

$$\text{Gross earnings} = \text{regular earnings} + \text{overtime earnings}$$

$$\text{Gross earnings} = \$480.00 + \$36.00$$

$$\text{Gross earnings} = \$516.00$$

(*b*) Estimate the annual gross earnings.

$$\text{Estimated annual earnings} = \begin{array}{c}\text{typical week's}\\ \text{gross earnings}\end{array} \times \begin{array}{c}\text{weeks paid for}\\ \text{in a year}\end{array}$$

$$\text{Estimated annual earnings} = \$516.00 \times 50$$

$$\text{Estimated annual earnings} = \$25{,}800$$

Salary

Management personnel and teachers are among the employees who are ordinarily hired for an annual salary. To calculate the amount of monthly earnings, divide the annual salary by the number of months in a year, 12. To calculate weekly earnings, divide the annual salary by the number of weeks in a year (52) or the number of weeks actually worked, whichever is appropriate for your purposes.

EXAMPLE Candy Ackridge is a teacher who earns a salary of $19,800 for a 9-month school year. She is paid, however, in 12 monthly payments. Calculate (*a*) the amount of Candy's gross earnings per paycheck received each of the 12 months in a year and (*b*) Candy's gross earnings for each of the 9 months actually worked.

Solution: (*a*) Calculate gross earnings per paycheck.

$$\text{Gross earnings per paycheck} = \begin{array}{c}\text{annual}\\ \text{salary}\end{array} \div \begin{array}{c}\text{months}\\ \text{in a year}\end{array}$$

$$\text{Gross earnings per paycheck} = \$19{,}800 \div 12$$

$$\text{Gross earnings per paycheck} = \$1{,}650$$

(*b*) Calculate gross earnings per month worked.

Gross earnings per month worked	=	annual salary	÷	number of months worked
Gross earnings per month worked	=	$19,800	÷	9
Gross earnings per month worked	=	$2,200		

Straight Commission

The earnings of many salespeople are calculated as a percent of the dollar amounts sold. To determine gross earnings, multiply the commission rate times the sales amount.

EXAMPLE Susan Ming is a real estate salesperson. She just sold a client's home for $240,000. Susan will receive a commission of $3\frac{1}{2}\%$ on the total sales price. Calculate her gross earnings from the sale.

Solution:

Gross earnings	=	sales amount	×	commission rate
Gross earnings	=	$240,000	×	$3\frac{1}{2}\%$
Gross earnings	=	$240,000	×	3.5%
Gross earnings	=	$8,400		

Salary Plus Commission

Some employees receive guaranteed salaries per week, month, or year and also commissions either on all sales made or on sales made above a certain amount called a *quota.*

When a commission is paid on all sales made, calculate the amount of commission by multiplying the commission rate by the sales amount, as illustrated in the preceding example. Gross earnings are calculated by adding the commissions earned to the salary.

When there is a sales quota that must be reached before a commission is earned, the amount of sales above the quota is first determined. Then the commission rate is multiplied by the sales amount above the quota to determine the commission earned. Gross earnings are calculated by adding the commissions earned to the salary. Of course, if sales are less than the quota, no commission is earned.

The two techniques described above are identical with those often followed to calculate the bonus earned by a manager.

EXAMPLE Howard Kinzel receives a weekly salary of $200 plus a 5% commission on all sales he makes above a $1,500 weekly quota. Last week he sold $3,600 of merchandise. Calculate Howard's gross earnings for the week.

Solution:

Step 1. Calculate sales made above the quota.

Sales above quota = total sales − quota
Sales above quota = $3,600 − $1,500
Sales above quota = $2,100

Step 2. Calculate the commission earned.

Commission earned = sales above quota × commission rate
Commission earned = $2,100 × 5%
Commission earned = $2,100 × .05
Commission earned = $105

Step 3. Calculate the week's gross earnings.

Gross earnings = salary + commission earned
Gross earnings = $200 + $105
Gross earnings = $305

Straight Piecework

Some employees who work for a manufacturer are paid a certain amount for each item produced, called *straight piecework.* To calculate gross earnings, multiply the number of pieces produced by the rate per piece.

EXAMPLE Jay Langly works for an electronics manufacturer and is paid $1.86 for each circuit board he solders. Yesterday he produced 33 boards. Calculate Jay's gross earnings for the day.

Solution:

Gross earnings = no. produced × rate each

Gross earnings = 33 × $1.86

Gross earnings = $61.38

Differential Piecework

To encourage employees to produce at high levels, some employers offer a *differential piecework* compensation plan. Under this arrangement, an employee earns an increasingly higher rate as the individual's production moves into higher brackets on the differential piecework scale.

The amount earned in each bracket of the scale is determined by multiplying the number of pieces produced by the pay rate for that bracket. Gross earnings are determined by adding the earnings in each of the scale's brackets.

EXAMPLE Karen Kilgore works as a seamstress in a clothing factory. She is paid on the following differential piecework scale for garments she produces per day. Yesterday, she produced 64 garments. Calculate her gross earnings.

DIFFERENTIAL PIECEWORK SCALE

Bracket	Items produced	Rate each
1	1 to 25	$ .80
2	26 to 50	.90
3	Over 50	1.00

Solution:

Bracket	No. produced	×	rate each	=	earnings per bracket
1	25	×	$ .80	=	$20.00
2	25	×	.90	=	22.50
3	14	×	1.00	=	14.00
	64				$56.50

DEDUCTIONS FROM EARNINGS

Employers are required to deduct federal income taxes and FICA taxes (Social Security) from their employees' gross earnings and to submit those amounts to the government. In addition, many states also impose an income tax which must be deducted from employee earnings and submitted to the state department of revenue.

Self-employed persons must submit quarterly payments for federal and state income taxes and must make FICA contributions as well.

Federal Income Tax

The amount of federal income tax due is determined by reference to tables like the partial ones shown on pages 123 and 124. Tables are available for daily, weekly, biweekly, semimonthly, monthly, and miscellaneous time periods. Different tables are provided for single and married taxpayers. The procedure for using all of the tables is the same. The tax rate is subject to change each year, and current tables can be acquired from your district internal revenue office (IRS) or from any accountant.

The amount of federal income tax withheld from your paycheck is determined by your gross earnings for the period and the number of *withholding allowances* (also called *exemptions*) you claim. You are allowed to claim one withholding allowance for yourself, one for each dependent, and, in some cases, allowances for special purposes as well.

EXAMPLE John Rivera is married and has two children. Last month his gross earnings were $2,540. He claims four withholding allowances. Calculate the amount of federal income taxes that should be withheld from his paycheck.

Solution:

Step 1. Use Table 7-1, "Married Persons—Monthly Payroll Period."

Step 2. Read down the two amount columns at the left under the headings "And the wages are—," "At least," "But less than" until you reach the amounts $2,520 under "At least" and $2,560 under "But less than." John's earnings of $2,540 fall between those two amounts.

Step 3. Read across to the right until you come to the amount shown in the column under the headings "And the number of withholding allowances claimed is—" and "4." The amount $348.80 will be withheld from John's monthly paycheck for federal income tax.

Taxpayers who wish to do so may declare fewer withholding allowances than they are entitled to. That will result in greater amounts being deducted from their paychecks, so that their take-home pay will be smaller. Then, unless they have outside income on which federal income tax has not been deducted, they most likely will be entitled to income tax refunds when they file their annual tax returns. Some persons who find it difficult to save money use that as a method of forced saving.

State Income Tax

Most, but not all, states impose a state income tax on wage earners. In those states in which a tax is assessed, the rate varies widely, from perhaps 5% to 20% of the amount of the federal income tax.

TABLE 7-1 FEDERAL INCOME TAX WITHHOLDING TABLE

SINGLE Persons—MONTHLY Payroll Period

(For Wages Paid After June 1983 and Before January 1985)

And the wages are—		And the number of withholding allowances claimed is—										
At least	But less than	0	1	2	3	4	5	6	7	8	9	10
		The amount of income tax to be withheld shall be—										
460	480	46.30	33.80	22.40	12.40	2.40	0	0	0	0	0	0
480	500	49.30	36.80	24.80	14.80	4.80	0	0	0	0	0	0
500	520	52.30	39.80	27.30	17.20	7.20	0	0	0	0	0	0
520	540	55.30	42.80	30.30	19.60	9.60	0	0	0	0	0	0
540	560	58.30	45.80	33.30	22.00	12.00	2.00	0	0	0	0	0
560	580	61.30	48.80	36.30	24.40	14.40	4.40	0	0	0	0	0
580	600	64.30	51.80	39.30	26.80	16.80	6.80	0	0	0	0	0
600	640	68.80	56.30	43.80	31.30	20.40	10.40	.40	0	0	0	0
640	680	74.80	62.30	49.80	37.30	25.20	15.20	5.20	0	0	0	0
680	720	80.80	68.30	55.80	43.30	30.80	20.00	10.00	0	0	0	0
720	760	86.80	74.30	61.80	49.30	36.80	24.80	14.80	4.80	0	0	0
760	800	92.80	80.30	67.80	55.30	42.80	30.30	19.60	9.60	0	0	0
800	840	99.90	86.30	73.80	61.30	48.80	36.30	24.40	14.40	4.40	0	0
840	880	107.50	92.30	79.80	67.30	54.80	42.30	29.80	19.20	9.20	0	0
880	920	115.10	99.30	85.80	73.30	60.80	48.30	35.80	24.00	14.00	4.00	0
920	960	122.70	106.90	91.80	79.30	66.80	54.30	41.80	29.30	18.80	8.80	0
960	1,000	130.30	114.50	98.60	85.30	72.80	60.30	47.80	35.30	23.60	13.60	3.60
1,000	1,040	137.90	122.10	106.20	91.30	78.80	66.30	53.80	41.30	28.80	18.40	8.40
1,040	1,080	145.50	129.70	113.80	98.00	84.80	72.30	59.80	47.30	34.80	23.20	13.20
1,080	1,120	153.10	137.30	121.40	105.60	90.80	78.30	65.80	53.30	40.80	28.30	18.00
1,120	1,160	160.70	144.90	129.00	113.20	97.40	84.30	71.80	59.30	46.80	34.30	22.80
1,160	1,200	168.30	152.50	136.60	120.80	105.00	90.30	77.80	65.30	52.80	40.30	27.80
1,200	1,240	177.10	160.10	144.20	128.40	112.60	96.70	83.80	71.30	58.80	46.30	33.80
1,240	1,280	187.10	167.70	151.80	136.00	120.20	104.30	89.80	77.30	64.80	52.30	39.80
1,280	1,320	197.10	176.30	159.40	143.60	127.80	111.90	96.10	83.30	70.80	58.30	45.80
1,320	1,360	207.10	186.30	167.00	151.20	135.40	119.50	103.70	89.30	76.80	64.30	51.80
1,360	1,400	217.10	196.30	175.40	158.80	143.00	127.10	111.30	95.50	82.80	70.30	57.80
1,400	1,440	227.10	206.30	185.40	166.40	150.60	134.70	118.90	103.10	88.80	76.30	63.80
1,440	1,480	237.10	216.30	195.40	174.60	158.20	142.30	126.50	110.70	94.80	82.30	69.80
1,480	1,520	247.10	226.30	205.40	184.60	165.80	149.90	134.10	118.30	102.40	88.30	75.80
1,520	1,560	257.10	236.30	215.40	194.60	173.80	157.50	141.70	125.90	110.00	94.30	81.80
1,560	1,600	267.10	246.30	225.40	204.60	183.80	165.10	149.30	133.50	117.60	101.80	87.80
1,600	1,640	277.10	256.30	235.40	214.60	193.80	172.90	156.90	141.10	125.20	109.40	93.80
1,640	1,680	287.10	266.30	245.40	224.60	203.80	182.90	164.50	148.70	132.80	117.00	101.20
1,680	1,720	297.10	276.30	255.40	234.60	213.80	192.90	172.10	156.30	140.40	124.60	108.80
1,720	1,760	307.10	286.30	265.40	244.60	223.80	202.90	182.10	163.90	148.00	132.20	116.40
1,760	1,800	317.10	296.30	275.40	254.60	233.80	212.90	192.10	171.50	155.60	139.80	124.00
1,800	1,840	327.10	306.30	285.40	264.60	243.80	222.90	202.10	181.30	163.20	147.40	131.60
1,840	1,880	338.40	316.30	295.40	274.60	253.80	232.90	212.10	191.30	170.80	155.00	139.20
1,880	1,920	350.40	326.30	305.40	284.60	263.80	242.90	222.10	201.30	180.40	162.60	146.80
1,920	1,960	362.40	337.40	315.40	294.60	273.80	252.90	232.10	211.30	190.40	170.20	154.40
1,960	2,000	374.40	349.40	325.40	304.60	283.80	262.90	242.10	221.30	200.40	179.60	162.00
2,000	2,040	386.40	361.40	336.40	314.60	293.80	272.90	252.10	231.30	210.40	189.60	169.60
2,040	2,080	398.40	373.40	348.40	324.60	303.80	282.90	262.10	241.30	220.40	199.60	178.80
2,080	2,120	410.40	385.40	360.40	335.40	313.80	292.90	272.10	251.30	230.40	209.60	188.80
2,120	2,160	422.40	397.40	372.40	347.40	323.80	302.90	282.10	261.30	240.40	219.60	198.80
2,160	2,200	434.40	409.40	384.40	359.40	334.40	312.90	292.10	271.30	250.40	229.60	208.80
2,200	2,240	446.40	421.40	396.40	371.40	346.40	322.90	302.10	281.30	260.40	239.60	218.80
2,240	2,280	458.40	433.40	408.40	383.40	358.40	333.40	312.10	291.30	270.40	249.60	228.80
2,280	2,320	470.40	445.40	420.40	395.40	370.40	345.40	322.10	301.30	280.40	259.60	238.80
2,320	2,360	483.40	457.40	432.40	407.40	382.40	357.40	332.40	311.30	290.40	269.60	248.80
2,360	2,400	497.00	469.40	444.40	419.40	394.40	369.40	344.40	321.30	300.40	279.60	258.80
2,400	2,440	510.60	482.20	456.40	431.40	406.40	381.40	356.40	331.40	310.40	289.60	268.80
2,440	2,480	524.20	495.80	468.40	443.40	418.40	393.40	368.40	343.40	320.40	299.60	278.80
2,480	2,520	537.80	509.40	481.10	455.40	430.40	405.40	380.40	355.40	330.40	309.60	288.80
2,520	2,560	551.40	523.00	494.70	467.40	442.40	417.40	392.40	367.40	342.40	319.60	298.80
2,560	2,600	565.00	536.60	508.30	480.00	454.40	429.40	404.40	379.40	354.40	329.60	308.80
2,600	2,640	578.60	550.20	521.90	493.60	466.40	441.40	416.40	391.40	366.40	341.40	318.80
2,640	2,680	592.20	563.80	535.50	507.20	478.80	453.40	428.40	403.40	378.40	353.40	328.80
2,680	2,720	605.80	577.40	549.10	520.80	492.40	465.40	440.40	415.40	390.40	365.40	340.40
2,720	2,760	619.40	591.00	562.70	534.40	506.00	477.70	452.40	427.40	402.40	377.40	352.40
2,760	2,800	633.60	604.60	576.30	548.00	519.60	491.30	464.40	439.40	414.40	389.40	364.40
2,800	2,840	648.40	618.20	589.90	561.60	533.20	504.90	476.60	451.40	426.40	401.40	376.40
2,840	2,880	663.20	632.40	603.50	575.20	546.80	518.50	490.20	463.40	438.40	413.40	388.40
2,880	2,920	678.00	647.20	617.10	588.80	560.40	532.10	503.80	475.40	450.40	425.40	400.40
		37 percent of the excess over $3,640 plus—										
$3.640 and over		951.80	921.00	890.10	859.30	828.50	797.60	766.80	736.00	705.10	674.30	643.50

TABLE 7-1 FEDERAL INCOME TAX WITHHOLDING TABLE (CONT.)

MARRIED Persons—MONTHLY Payroll Period

(For Wages Paid After June 1983 and Before January 1985)

And the wages are—		And the number of withholding allowances claimed is—										
At least	But less than	0	1	2	3	4	5	6	7	8	9	10
		The amount of income tax to be withheld shall be—										
480	500	34.80	24.80	14.80	4.80	0	0	0	0	0	0	0
500	520	37.20	27.20	17.20	7.20	0	0	0	0	0	0	0
520	540	39.60	29.60	19.60	9.60	0	0	0	0	0	0	0
540	560	42.00	32.00	22.00	12.00	2.00	0	0	0	0	0	0
560	580	44.40	34.40	24.40	14.40	4.40	0	0	0	0	0	0
580	600	46.80	36.80	26.80	16.80	6.80	0	0	0	0	0	0
600	640	50.40	40.40	30.40	20.40	10.40	.40	0	0	0	0	0
640	680	55.20	45.20	35.20	25.20	15.20	5.20	0	0	0	0	0
680	720	60.00	50.00	40.00	30.00	20.00	10.00	0	0	0	0	0
720	760	64.80	54.80	44.80	34.80	24.80	14.80	4.80	0	0	0	0
760	800	69.60	59.60	49.60	39.60	29.60	19.60	9.60	0	0	0	0
800	840	75.40	64.40	54.40	44.40	34.40	24.40	14.40	4.40	0	0	0
840	880	82.20	69.20	59.20	49.20	39.20	29.20	19.20	9.20	0	0	0
880	920	89.00	74.80	64.00	54.00	44.00	34.00	24.00	14.00	4.00	0	0
920	960	95.80	81.60	68.80	58.80	48.80	38.80	28.80	18.80	8.80	0	0
960	1,000	102.60	88.40	74.30	63.60	53.60	43.60	33.60	23.60	13.60	3.60	0
1,000	1,040	109.40	95.20	81.10	68.40	58.40	48.40	38.40	28.40	18.40	8.40	0
1,040	1,080	116.20	102.00	87.90	73.70	63.20	53.20	43.20	33.20	23.20	13.20	3.20
1,080	1,120	123.00	108.80	94.70	80.50	68.00	58.00	48.00	38.00	28.00	18.00	8.00
1,120	1,160	129.80	115.60	101.50	87.30	73.10	62.80	52.80	42.80	32.80	22.80	12.80
1,160	1,200	136.60	122.40	108.30	94.10	79.90	67.60	57.60	47.60	37.60	27.60	17.60
1,200	1,240	143.40	129.20	115.10	100.90	86.70	72.60	62.40	52.40	42.40	32.40	22.40
1,240	1,280	150.20	136.00	121.90	107.70	93.50	79.40	67.20	57.20	47.20	37.20	27.20
1,280	1,320	157.00	142.80	128.70	114.50	100.30	86.20	72.00	62.00	52.00	42.00	32.00
1,320	1,360	163.80	149.60	135.50	121.30	107.10	93.00	78.80	66.80	56.80	46.80	36.80
1,360	1,400	170.60	156.40	142.30	128.10	113.90	99.80	85.60	71.60	61.60	51.60	41.60
1,400	1,440	177.40	163.20	149.10	134.90	120.70	106.60	92.40	78.20	66.40	56.40	46.40
1,440	1,480	184.20	170.00	155.90	141.70	127.50	113.40	99.20	85.00	71.20	61.20	51.20
1,480	1,520	191.00	176.80	162.70	148.50	134.30	120.20	106.00	91.80	77.70	66.00	56.00
1,520	1,560	197.80	183.60	169.50	155.30	141.10	127.00	112.80	98.60	84.50	70.80	60.80
1,560	1,600	204.60	190.40	176.30	162.10	147.90	133.80	119.60	105.40	91.30	77.10	65.60
1,600	1,640	212.50	197.20	183.10	168.90	154.70	140.60	126.40	112.20	98.10	83.90	70.40
1,640	1,680	221.30	204.00	189.90	175.70	161.50	147.40	133.20	119.00	104.90	90.70	76.50
1,680	1,720	230.10	211.80	196.70	182.50	168.30	154.20	140.00	125.80	111.70	97.50	83.30
1,720	1,760	238.90	220.60	203.50	189.30	175.10	161.00	146.80	132.60	118.50	104.30	90.10
1,760	1,800	247.70	229.40	211.00	196.10	181.90	167.80	153.60	139.40	125.30	111.10	96.90
1,800	1,840	256.50	238.20	219.80	202.90	188.70	174.60	160.40	146.20	132.10	117.90	103.70
1,840	1,880	265.30	247.00	228.60	210.30	195.50	181.40	167.20	153.00	138.90	124.70	110.50
1,880	1,920	274.10	255.80	237.40	219.10	202.30	188.20	174.00	159.80	145.70	131.50	117.30
1,920	1,960	282.90	264.60	246.20	227.90	209.60	195.00	180.80	166.60	152.50	138.30	124.10
1,960	2,000	292.10	273.40	255.00	236.70	218.40	201.80	187.60	173.40	159.30	145.10	130.90
2,000	2,040	302.10	282.20	263.80	245.50	227.20	208.80	194.40	180.20	166.10	151.90	137.70
2,040	2,080	312.10	291.30	272.60	254.30	236.00	217.60	201.20	187.00	172.90	158.70	144.50
2,080	2,120	322.10	301.30	281.40	263.10	244.80	226.40	208.10	193.80	179.70	165.50	151.30
2,120	2,160	332.10	311.30	290.40	271.90	253.60	235.20	216.90	200.60	186.50	172.30	158.10
2,160	2,200	342.10	321.30	300.40	280.70	262.40	244.00	225.70	207.40	193.30	179.10	164.90
2,200	2,240	352.10	331.30	310.40	289.60	271.20	252.80	234.50	216.20	200.10	185.90	171.70
2,240	2,280	362.10	341.30	320.40	299.60	280.00	261.60	243.30	225.00	206.90	192.70	178.50
2,280	2,320	372.10	351.30	330.40	309.60	288.80	270.40	252.10	233.80	215.40	199.50	185.30
2,320	2,360	382.10	361.30	340.40	319.60	298.80	279.20	260.90	242.60	224.20	206.30	192.10
2,360	2,400	392.10	371.30	350.40	329.60	308.80	288.00	269.70	251.40	233.00	214.70	198.90
2,400	2,440	402.40	381.30	360.40	339.60	318.80	297.90	278.50	260.20	241.80	223.50	205.70
2,440	2,480	413.60	391.30	370.40	349.60	328.80	307.90	287.30	269.00	250.60	232.30	214.00
2,480	2,520	424.80	401.50	380.40	359.60	338.80	317.90	297.10	277.80	259.40	241.10	222.80
2,520	2,560	436.00	412.70	390.40	369.60	348.80	327.90	307.10	286.60	268.20	249.90	231.60
2,560	2,600	447.20	423.90	400.60	379.60	358.80	337.90	317.10	296.30	277.00	258.70	240.40
2,600	2,640	458.40	435.10	411.80	389.60	368.80	347.90	327.10	306.30	285.80	267.50	249.20
2,640	2,680	469.60	446.30	423.00	399.60	378.80	357.90	337.10	316.30	295.40	276.30	258.00
2,680	2,720	480.80	457.50	434.20	410.80	388.80	367.90	347.10	326.30	305.40	285.10	266.80
2,720	2,760	492.00	468.70	445.40	422.00	398.80	377.90	357.10	336.30	315.40	294.60	275.60
2,760	2,800	503.20	479.90	456.60	433.20	409.90	387.90	367.10	346.30	325.40	304.60	284.40
2,800	2,840	514.40	491.10	467.80	444.40	421.10	397.90	377.10	356.30	335.40	314.60	293.80
2,840	2,880	526.10	502.30	479.00	455.60	432.30	409.00	387.10	366.30	345.40	324.60	303.80
2,880	2,920	539.30	513.50	490.20	466.80	443.50	420.20	397.10	376.30	355.40	334.60	313.80
2,920	2,960	552.50	525.00	501.40	478.00	454.70	431.40	408.00	386.30	365.40	344.60	323.80
		37 percent of the excess over $4,600 plus—										
$4,600 and over		1,135.00	1,104.20	1,073.30	1,042.50	1,011.70	980.80	950.00	919.20	888.30	857.50	826.70

In many states, a tax table that is similar to the federal income tax withholding tables is used to calculate the tax. In other states, the tax is simply a certain percent of the federal income tax. In most states, the same tax table or calculation procedure is used for both single and married persons. Contact your state department of revenue or an accountant for a current income tax table or calculation procedure.

FICA Tax

The Federal Insurance Contributions Act tax (FICA) is commonly called Social Security. Employers calculate the amount of tax to be withheld from employee earnings by referring to a table that is similar to the federal income tax withholding tables or by multiplying the employee's eligible earnings by the FICA rate.

Employers must match the amounts deducted from employee earnings and contribute that amount to the FICA fund. Self-employed persons are taxed at a higher rate than are persons who are employees, since no matching contribution is made by an employer.

Table 7-2 supplies the most currently available information for the percent of contribution that applies to earnings of employees and self-employed persons. It also shows the maximum earnings to which the tax applies for 1983, which is $35,700. In other words, the FICA tax must be paid on the first $35,700 earned in 1983. At the time of writing, maximum earnings for subsequent years had not yet been set.

If a person works as an employee for two or more employers during a year, each employer will deduct the FICA tax from the employee's earnings until the maximum earnings have been reached at that job. An employee whose total earnings from various jobs exceed the maximum earnings amount will receive a refund for overpaid FICA tax when the federal income tax return is filed.

An employee who has earnings of less than the maximum earnings amount pays the FICA tax on the total earnings. If that person is also self-employed at another endeavor, he or she must make an additional FICA contribution at the self-employed rate until the total earnings reach the maximum earnings amount.

TABLE 7-2 FICA TAX CONTRIBUTIONS

Year	Maximum earnings	Employee's contribution	Self-employed person's contribution
1983	$35,700	6.70%	9.35%
1984	?	6.70%*	11.13%*
1985	?	7.05%	11.80%*
1986	?	7.15%	12.30%*
1987	?	7.15%	12.30%*
1988	?	7.51%	15.02%
1989	?	7.51%	15.02%
1990	?	7.65%	15.30%

*The officially published FICA tax rate is higher than that shown, but a credit reduces the effective, or actual, rate to that presented here.

It should be pointed out that even though the percent of contribution for employees and self-employed persons is shown for 1984 to 1990, the rates are subject to revision. In the past, the estimated rates on tables such as this have often been increased.

EXAMPLE 1 Millie Uhlenkamp, an employee at Mason Company, had gross earnings of $2,460 in March, all of which is subject to FICA tax. The tax rate is 6.7%. Calculate the amount deducted from Millie's gross earnings for FICA tax.

Solution:

FICA tax = eligible earnings × employee's FICA tax rate

FICA tax = $2,460 × 6.7%

FICA tax = $2,460 × .067

FICA tax = $164.82

EXAMPLE 2 Marshall Edmunds was employed at Canton Company, where his gross earnings for the year 1983 were $31,500. He also earned $14,500 as a self-employed musician. Maximum annual earnings upon which FICA tax must be paid were $35,700 in 1983, and the FICA tax rate was 6.7% for employees and 9.35% for self-employed persons. Calculate (*a*) the amount of FICA tax deducted from Marshall's earnings at Canton Company, (*b*) the amount of FICA tax Marshall paid based on his income as a musician, and (*c*) Marshall's total FICA contribution for the year.

Solution: (*a*) Calculate the FICA tax attributed to Canton Co. employment.

$$\text{FICA tax} = \text{eligible earnings} \times \text{employee's FICA tax rate}$$

$$\text{FICA tax} = \$31{,}500 \times 6.7\%$$

$$\text{FICA tax} = \$31{,}500 \times .067$$

$$\text{FICA tax} = \$2{,}110.50$$

(*b*) Calculate the FICA tax attributed to self-employed income.

Step 1. Calculate the amount of earnings subject to self-employed FICA tax rate.

$$\text{Self-employed taxable earnings} = \begin{matrix}\text{FICA}\\ \text{maximum}\\ \text{earnings}\end{matrix} - \begin{matrix}\text{earnings previously}\\ \text{taxed}\end{matrix}$$

$$\text{Self-employed taxable earnings} = \$35{,}700 - \$31{,}500$$

$$\text{Self-employed taxable earnings} = \$4{,}200$$

Step 2. Calculate the FICA tax on self-employed earnings.

$$\text{Self-employed FICA tax} = \begin{matrix}\text{self-employed}\\ \text{taxable earnings}\end{matrix} \times \begin{matrix}\text{self-employed}\\ \text{FICA rate}\end{matrix}$$

$$\text{Self-employed FICA tax} = \$4{,}200 \times 9.35\%$$

$$\text{Self-employed FICA tax} = \$4{,}200 \times .0935$$

$$\text{Self-employed FICA tax} = \$392.70$$

(*c*) Calculate year's total FICA tax.

$$\text{Total FICA tax} = \begin{matrix}\text{employee's}\\ \text{FICA tax}\end{matrix} + \begin{matrix}\text{self-employed}\\ \text{FICA tax}\end{matrix}$$

$$\text{Total FICA tax} = \$2{,}110.50 + \$392.70$$

$$\text{Total FICA tax} = \$2{,}503.20$$

Calculating Take-Home Pay

An employee's take-home pay (also called *net pay*) is the amount which remains after federal income tax, state income tax, FICA tax, and voluntary deductions are subtracted from gross earnings.

EXAMPLE Gary Thompson is employed at Wickert Company. His gross earnings for September were $2,820. Deductions from his gross earnings were as follows: federal income tax, $467.80; state income tax, $81.85; FICA tax, $188.94, and union dues, $35.00. Calculate Gary's net pay for the month.

Solution:

$$\text{Net pay} = \begin{array}{c}\text{gross}\\\text{earnings}\end{array} - \left(\begin{array}{c}\text{federal}\\\text{inc. tax}\end{array} + \begin{array}{c}\text{state}\\\text{inc. tax}\end{array} + \begin{array}{c}\text{FICA}\\\text{tax}\end{array} + \begin{array}{c}\text{union}\\\text{dues}\end{array}\right)$$

$$\text{Net pay} = \$2{,}820 - (\$467.80 + \$81.85 + \$188.94 + \$35.00)$$

$$\text{Net pay} = \$2{,}820 - \$773.59$$

$$\text{Net pay} = \$2{,}046.41$$

ANALYZING THE REAL INCOME FROM OVERTIME PAY OR A SECOND JOB

Is it worth working overtime or taking a second job to earn an additional $300, $400, $500, or more per month? Some people think it is; others guess it is not. Well, the only way to be certain is to calculate what the net result of earning that extra money will be. In other words, the net income received from the extra earnings after subtracting the related deductions for federal and state income tax, FICA tax, and additional expenses connected with taking that job must be calculated.

If the extra income from overtime pay or a second job is substantial, it will most likely boost your total earnings into a higher federal income tax bracket, and probably into a higher state income tax bracket as well. Also, if you have not already reached the FICA maximum earnings limit, you will have additional FICA taxes to pay.

If you take a second job where you earn, say, $400 a month, the amount deducted from your paycheck for federal and state income taxes probably will be low. That is because your employer will use the $400 amount when applying the income tax withholding tables and the tax on that small amount of income, when considered by itself, is low. In reality, however, when you file your income tax return at the end of the year, the extra $400 income will be *added* to your regular income and a much higher tax rate will apply. Therefore, either you may need to pay additional taxes at that time or your refund will be smaller. Thus, when you are analyzing the effect of earning additional income, add the extra income to your regular income when you calculate the amount of the taxes that apply.

EXAMPLE John Olson is employed by Stover Company, where his gross earnings are $2,000 per month ($24,000 per year). He claims two withholding allowances. Deductions are made for federal income tax, state income tax, and FICA tax.

John has just been offered a part-time job as a retail clerk in a shopping center, where he will gross $500 per month for 100 hours of work ($5 per hour). John estimates that he will incur an additional $75 per month in expenses for auto use, clothing, and so on, if he takes the part-time job.

Calculate (*a*) John's current net pay per month, based on his job at Stover Company, (*b*) John's net pay per month that would be received from working at both jobs, (*c*) John's additional net pay per month that would be received from working at the part-time job, and (*d*) John's additional net income per month, after deducting related expenses, which would result from taking the part-time job.

Use the excerpt from the federal income tax withholding tables headed "Married Persons—Monthly Payroll Period." Calculate the state income tax to be 15% of the federal income tax amount. The FICA tax rate is 6.7%, based on maximum annual earnings of $35,700.

Solution: (*a*) Calculate John's current monthly net income from Stover Company. [Reference to the federal income tax withholding table shows that $263.80 is deducted each month for federal income tax. The state income tax is $39.57 ($263.80 × 15% = $39.57). The FICA tax is $134.00 ($2,000 × 6.7% = $134.00).]

$$\text{Net pay} = \text{gross earnings} - \left(\text{federal inc. tax} + \text{state inc. tax} + \text{FICA taxes}\right)$$

$$\text{Net pay} = \$2{,}000.00 - (\$263.80 + \$39.57 + \$134.00)$$

$$\text{Net pay} = \$2{,}000.00 - \$437.37$$

$$\text{Net pay} = \$1{,}562.63$$

(*b*) Calculate the net pay John would receive from working at both jobs. [The total gross earnings per month is $2,500. Reference to the federal income tax withholding table shows that $380.40 would be the federal income tax per month. The state income tax would be $57.06 ($380.40 × 15% = $57.06). Since John's combined income from both jobs would be below the FICA maximum earnings limit of $35,700, his total monthly earnings would be subject to FICA tax. The FICA tax would be $167.50 per month ($2,500 × 6.7% = $167.50).]

$$\text{Net pay both jobs} = \text{gross earnings} - \left(\text{federal inc. tax} + \text{state inc. tax} + \text{FICA taxes}\right)$$

$$\text{Net pay both jobs} = \$2{,}500.00 - (\$380.40 + \$57.06 + \$167.50)$$

$$\text{Net pay both jobs} = \$2{,}500.00 \qquad \$604.96$$

$$\text{Net pay both jobs} = \$1{,}895.04$$

(*c*) Calculate the additional net pay that would be earned per month from taking the part-time job.

$$\text{Additional net pay} = \text{net pay, both jobs} - \text{current net pay (from Stover Co.)}$$

$$\text{Additional net pay} = \$1{,}895.04 - \$1{,}562.63$$

$$\text{Additional net pay} = \$332.41$$

(*d*) Calculate the net income John would earn from the part-time job after deducting additional expenses related to the extra job.

$$\text{Part-time job net income} = \frac{\text{additional}}{\text{net pay}} - \frac{\text{job-related}}{\text{expenses}}$$

$$\text{Part-time job net income} = \$332.41 - \$75.00$$

$$\text{Part-time job net income} = \$257.41$$

We can see from the preceding analysis that even though John can increase his gross earnings by $500 per month by taking the part-time job, he will actually receive only $257.41 after taxes and expenses ($2.57 per hour). Should John take the job? Only John can decide.

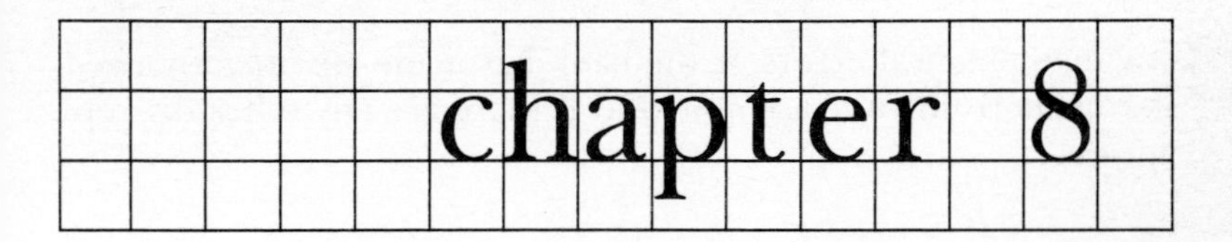

Checking Account Calculations

A checking account can be a safe, easy, and convenient way to transfer funds and have a receipt of payment for your records at the same time. It can also be a source of exasperation, particularly when your checkbook balance doesn't match the bank's records. Then too, the many different interest-bearing checking account plans offered by various financial institutions can be a source of bewilderment: just which plan will pay you the most interest anyway?

This section is designed to help you eliminate those exasperations and bewilderments, assist you in verifying the accuracy of your checking account records, and provide a method for determining which interest-bearing checking account plan is best for you.

RECONCILING YOUR BANK STATEMENT

The term "reconciling your bank statement" means to compare your checkbook record (check stub or check register) with the bank's record of your account (the bank statement). The purpose of the comparison is to determine if you have made or the bank has made any mathematical errors, if the bank has made any unauthorized payments from your account, and if there are any other discrepancies.

Starting from the Beginning

To perform an accurate bank reconciliation, the process must be started from the time your account was first opened and your first bank statement was received. Therefore, if your checking account has been in use for some time and you have never completed a bank reconciliation, you must go back to the very beginning and work your way through to the most recent bank statement received.

If your checking account has been in use for a lengthy period, say, several years, reconciliation from the beginning can be a mountainous task. Therefore, if you happen to be in that situation, a simpler solution is suggested: stop using your current checking account and open a new one. Then from the time your first bank statement is received for your new account, perform a bank reconciliation and continue doing so month after month. After all checks and transactions on your original checking account have cleared the bank, close that account and transfer the balance to your new account. You might find from studying this book's section titled "Selecting the Best Interest-Bearing Checking Account" that it may be wise to select a different bank's checking account plan than the one you are currently using anyway.

Gathering Bank Reconciliation Data

When you receive the bank statement, you should compare it with your checkbook record. Ordinarily, if yours is an active checking account to which you frequently make deposits and on which you write checks, the ending balance shown on the bank statement will not be the same as your checkbook balance on the same date. One reason is that some checks written by you, and deducted from your checkbook record, may not have yet been presented to your bank for payment. (They are called *outstanding checks.*)

Likewise, if you mailed a deposit to the bank shortly before the bank statement date (called a *deposit in transit*), it will be shown on your checkbook record but not on the bank statement. Similarly, the bank statement might show an addition for interest earned and/or a deduction for service charges, which will not appear on your checkbook record. There may be other differences because of automatic deposits made to your account, automatic payments made from your account, and so on. When you gather data for preparing your bank reconciliation, compare bank statement amounts with your checkbook to locate any amounts shown on one of the records but not on the other.

Since many bank statements now list checks in order by check number, it is easy to identify outstanding checks. If your bank does not provide that service, arrange the canceled checks received with your bank statement in numerical order. In either case, the dollar amount of any outstanding checks can then be determined by reference to your checkbook records. Even if the bank statement does list canceled checks in order, review all of your canceled checks to verify that there are no unauthorized checks or forgeries.

Bank Reconciliation Process

A form like that shown in Table 8-1 on page 138 is used to perform the bank reconciliation. The first step is to list the ending balance shown on your checkbook record and the ending balance shown on the bank statement as of the bank statement date. Additions and subtractions are then made to these ending balances as follows. Of course, your bank reconciliation may not contain all of these adjustments.

Adjustments to the Ending Checkbook Balance

Add: Interest earned on the account
Automatic deposits made to your bank account

Subtract: Service charges assessed by the bank
Automatic payments made by the bank

Adjustments to the Ending Bank Statement Balance

Add: Deposits in transit

Subtract: Outstanding checks

If the adjusted checkbook balance and the adjusted bank statement balance are equal, you can assume that there are no errors in your records or in the bank's records and that there are no other discrepancies. On the other hand, if the two adjusted balances are not equal, an error has been made somewhere and you must search until you find it.

Banks rarely make errors in maintaining their depositor's accounts (although it can happen). Therefore, your first step in locating the error should be to recheck your own calculations, which can be simplified by these steps:

1. Recheck your addition and subtraction in completing the bank reconciliation form.
2. Carefully compare the bank statement with your checkbook record to be certain that all amounts that should be included on the bank reconciliation have been identified.
3. Recheck the addition and subtraction in your checkbook record since the last bank reconciliation was performed.
4. If you still cannot locate the error, after rechecking your work several times, you might ask a knowledgeable friend to review your work, since you might be consistently overlooking the same error.
5. If you do not locate the error through this process, check the bank statement calculations to determine if the bank has made an error.
6. If you still cannot locate the error, you might consider asking an employee of your bank to assist you.

Keep in mind that if you fail to get your bank reconciliation form to balance, all future bank reconciliations will most likely also fail to balance! Therefore, keep working at it until you locate the discrepancies. On the bright side, after you have performed one or two bank reconciliations, you will become very familiar with the process, which will become increasingly easy to complete.

EXAMPLE Sarah Carron received her bank statement from National Bank covering the period September 1 to 30. It shows that the ending balance (September 30) was $728.60, that $2.15 interest was earned on the account, and that a service charge of $4.50 was assessed. Also, an automatic payment of $38.50 was made by the bank to New York Life Insurance Company for Sarah's insurance premium. Sarah had not deducted the insurance premium from her checkbook record.

Sarah's checkbook balance on September 30 was $639.30. Upon comparing her checkbook record with the bank statement, Sarah discovered that the following checks written on or before September 30 were outstanding: No. 302, $97.80; No. 316, $45.10; and No. 319, $12.25; (the outstanding checks total $155.15). Also, a deposit for $25.00, mailed to the bank on September 29, was not shown on the bank statement; it is recorded on Sarah's checkbook record.

Prepare a bank reconciliation for Sarah by using Table 8-1.

The adjusted checkbook balance and the adjusted bank statement balance are equal ($598.45). Therefore, we can assume no errors are present in Sarah's checkbook records or in the bank's records.

Updating the Checkbook Record

On the top portion of the bank reconciliation form are recorded several additions and subtractions that were made to the checking account on the bank's records but do not appear on the checkbook record. Therefore, those changes must now be made on the checkbook record (the check stub or check register).

Specifically, in the preceding example, Sarah Carron adds $2.15 to her checkbook balance for interest earned and subtracts the $4.50 service charge and $38.50 life insurance premium while properly labeling each amount. That having been done, the checkbook record is up to date.

Solution:

TABLE 8-1 BANK RECONCILIATION

a. *Ending Checkbook Balance:*		$ 639.30
Add:		
INTEREST EARNED	$ 2.15	
b. Total Additions:	$ 2.15	
c. Subtotal (a + b):		$ 641.45
Subtract:		
SERVICE CHARGE	$ 4.50	
N.Y. LIFE INSURANCE PREMIUM	38.50	
d. Total Subtractions:	$ 43.00	
e. Adjusted Checkbook Balance (c − d):		$ 598.45
f. *Ending Bank Statement Balance:*		$ 728.60
Add:		
DEPOSIT IN TRANSIT	$ 25.00	
g. Total Additions:	$ 25.00	
h. Subtotal (f + g)		$ 753.60
Subtract:		
OUTSTANDING CHECKS (TOTAL)	$ 155.15	
i. Total Subtractions:	$ 155.15	
j. Adjusted Bank Statement Balance (h − i):		$ 598.45

SELECTING THE BEST INTEREST-BEARING CHECKING ACCOUNT

All banks, savings and loan associations, and credit unions, as well as other types of financial institutions, now offer interest-bearing checking accounts of some kind. Even though these accounts operate similarly, there are marked differences in the way interest is calculated and how service charges are assessed.

Interest Calculation Methods

The following plans are representative of those used to calculate interest on checking accounts:

1. Simple interest is paid on the average of the daily ending balances or on each day's ending balance in the account.
2. Simple interest is paid by considering the lowest ending balance on any day during the statement period to be the average daily ending balance.
3. Interest is compounded on the average of the daily ending balances or on each day's ending balance.

Service Charge Plans

The following plans are representative of those used to calculate the service charge on checking accounts:

1. No service charge is assessed for checks written.
2. No service charge is assessed if the account has at least a balance of $500, $300, or some other stated amount *at all times.*
3. No service charge is assessed if the *average* daily ending balance in the account is $1,000, $500, or some other amount.
4. A service charge is assessed all customers for all checks written.

When checking account service charges are assessed, it is usually at a flat rate of, say, 10¢ to 20¢ for each check processed by the bank. In addition, many financial institutions require that their depositors buy check blanks, regardless of whether they qualify for service-charge-free checking.

Even if a service charge is assessed, depositors still earn interest on their accounts, and the interest earned will partially or fully offset the service charges and the cost of buying checks.

Analyzing Your Checking Account

Which interest-bearing checking account will provide you with the highest interest earned and the lowest service charge? Well, that depends. Mostly, it depends on the minimum balance or average daily ending balance you maintain in your account and on the number of checks you write.

The first step, therefore, in determining which interest-bearing checking account will be the most beneficial for you is to analyze your current checking account data. In gathering your data, select a recent bank statement period (usually a month) that is representative of how your account normally operates as to the number of checks written and daily ending account balances. If no single bank statement seems to be representative of your normal month-to-month checking account activities, average the information from two or more bank statements. If you do not currently have a checking account but are considering opening one, estimate the data as closely as possible.

Determining the Number of Checks Written. Many bank statements list the number of checks processed during the month, and you might simply be able to obtain the number from that source. Otherwise, count the canceled checks included with the bank statement or obtain the number from your checkbook record.

Determining Your Account's Lowest Ending Daily Balance. Here the goal is to determine the lowest ending balance in your account on any day during the month. Simply scan the "balance" column of your bank statement to locate the lowest balance shown. For instance, the lowest ending daily balance in the bank statement information in the following example is $71.00, on June 19.

Determining the Average Daily Ending Balance. The term "average daily ending balance" means what it implies: it is the average of the ending balances on the days that make up the statement period. Many bank statements have this amount calculated and listed for you. If the average daily ending balance is not shown on your current bank statement, you must calculate it as follows:

1. List the ending balance for each day of the statement period. (If no checks are processed and no deposits are recorded on a particular day, the bank statement will not show a balance for that day. Therefore, that day's ending balance is the same as the immediately preceding day's balance that is listed.)
2. Divide the total daily ending balances (as calculated in step 1) by the number of days in the statement period.

EXAMPLE Information taken from Romey Snyder's bank statement for June 1 to 30 is shown below. Calculate the average daily ending balance.

Date	Add deposits	Deduct checks	Ending balance
June 1			$216.00
4		$ 20.00	196.00
6	$800.00		996.00
7		50.00	946.00
13		200.00	746.00
17		500.00	246.00
19		175.00	71.00
20	800.00		871.00
24		360.00	511.00
30		275.00	236.00

Solution:

Step 1. Determine the total daily balances. See Table 8-2 on page 142.

Step 2. Calculate the average daily ending balance.

$$\text{Average daily ending balance} = \text{total daily ending balances} \div \text{number of days in statement period}$$

$$\text{Average daily ending balance} = \$18{,}045.00 \div 30$$

$$\text{Average daily ending balance} = \$601.50$$

TABLE 8-2 TOTAL DAILY BALANCES

Date	Ending balance	Date	Ending balance
June 1	$ 216.00	June 17	$ 246.00
2	216.00	18	246.00
3	216.00	19	71.00
4	196.00	20	871.00
5	196.00	21	871.00
6	996.00	22	871.00
7	946.00	23	871.00
8	946.00	24	511.00
9	946.00	25	511.00
10	946.00	26	511.00
11	946.00	27	511.00
12	946.00	28	511.00
13	746.00	29	511.00
14	746.00	30	236.00
15	746.00	Total	$18,045.00
16	746.00		

Calculating Interest Earned on Your Account

Interest on checking accounts is calculated as being either simple interest or interest compounded daily. (A third method, interest compounded continually, also is used, but the result is very similar to that obtained by compounding daily, and it is therefore not included here.)

As you will see from the following examples, the amount of interest earned on a checking account with a relatively small balance varies little with a small change in interest rate or when one interest calculation method instead of another is used.

Calculating Simple Interest. When simple interest is calculated, the following formula is used:

$$\text{Interest} = \text{principal} \times \text{rate} \times \text{time}$$

Principal refers to the balance of your account, on which interest is paid. *Rate* refers to the percent of interest paid. *Time* refers to the length of time in the statement period. Since interest rates are stated in terms of a year, the "time" element of the formula is expressed as a fraction with the number of days in the statement period as the numerator (top number) and the number of days in a year, 365, as the denominator (bottom number).

EXAMPLE The average daily ending balance in Janet Sporer's checking account for the 30-day bank statement period June 1 to June 30 was $500. The bank pays simple interest at the rate of 5.5% on the average daily ending balance. Calculate the amount of interest earned on the account.

Solution:

$$\text{Interest earned} = \text{principal} \times \text{rate} \times \text{time}$$

$$\text{Interest earned} = \$500 \times 5.5\% \times \frac{30}{365}$$

$$\text{Interest earned} = \$500 \times .055 \times (30 \div 365)$$

$$\text{Interest earned} = \$500 \times .055 \times .0822$$

$$\text{Interest earned} = \$2.26$$

Calculating Interest Compounded Daily. Banks ordinarily use a computer when calculating interest compounded daily. Another method, which will yield the same results, is to use a table like that shown here. Table 8-3 gives the amount of interest earned on a principal amount of $1 for a 30- or 31-day period, the normal length of a bank statement period.

TABLE 8-3 INTEREST EARNED ON $1.00 WHEN COMPOUNDED DAILY

	Stated interest rate				
Days	5.00%	5.25%	5.50%	5.75%	6.00%
30	.0041178	.0043241	.0045304	.0047368	.0049433
31	.0042553	.0044685	.0046818	.0048951	.0051085

To calculate the amount of interest earned on an account when interest is compounded daily, simply multiply the balance of the account on which interest is paid (the principal) by the appropriate amount selected from Table 8-3.

EXAMPLE The average daily ending balance in Daniel Allen's checking account for the 30-day bank statement period June 1 to June 30 was $500. The bank pays interest compounded at 5.50% on the average daily ending balances. Calculate the amount of interest earned on the account by using the table of interest earned on $1.00 shown on page 143.

Solution:

Interest earned	=	principal	×	interest compounded daily on $1.00 at 5.50% for 30 days
Interest earned	=	$500	×	.0045304
Interest earned	=	$2.27		

Comparing Interest-Bearing Checking Accounts

The goal in comparing various interest-bearing checking account plans is to be able to select the one that will provide you with the best net interest income and net service charge position. That is the plan that will give you the highest interest income after deducting service charges or the lowest service charge after taking interest earned on your account into consideration. In all calculations, the cost of purchasing check blanks must also be included.

The process for comparing various interest-bearing checking accounts is as follows:

1. Gather the following information about your current checking account by using the procedure described in the preceding pages:
   a. Number of checks written in a typical month
   b. Lowest daily ending balance on your bank statement during a typical month
   c. Average daily ending balance on your bank statement during a typical month

2. From each financial institution in your area, secure a written statement which describes the specifics of its interest-bearing checking accounts. Be sure to gather the following information:
   a. Rate of interest paid on the account balance
   b. Whether simple interest or interest compounded daily (or continuously) is calculated
   c. Basis upon which interest is calculated, such as the lowest day's ending balance, the average daily ending balance, or each day's ending balance
   d. Circumstances in which a bank service charge is or is not charged, such as there being no service charge for processing checks if a minimum balance is maintained in the account
   e. Method of calculating service charges, such as cost per check processed, if a service charge is assessed
   f. Cost, if any, of buying check blanks
3. After all of that information is gathered, calculate the following for each financial institution's interest-bearing checking account plan. Use your own data for a typical month, as identified in step 1.
   a. Interest earned on the account
   b. Service charge assessed, if any
   c. Cost, if any, of check blanks
   d. The net difference between interest income earned and the service charges and the cost of check blanks

The results should clearly indicate which interest-bearing checking account is most beneficial for you.

EXAMPLE Jim and Beth Sammons gathered the following information, which represents a typical month's activity in their checking account:

Number of checks written	40
Lowest daily ending balance	$150
Average daily ending balance	$600

Jim and Beth gathered the following information about interest-bearing checking accounts available in their community.

(*a*) Franklin National Bank

1. Rate of interest paid: 5.25%, compounded on the average daily ending balance.
2. If an average daily ending balance of $500 or more is maintained, there is no service charge. If the average daily ending balance is less than $500, a 15¢ service charge is assessed for each check processed.
3. All depositors must purchase check blanks. The cost is $4.00 for 100 checks (4¢ per check).

(*b*) Martin County Savings and Loan Association

1. Rate of interest paid: 5.50% simple interest, paid on the average daily ending balance.
2. If an average daily ending balance of $1,000 or more is maintained, there is no service charge. If the average daily ending balance is less than $1,000, a 15¢ service charge is assessed for each check processed.
3. There is no charge for check blanks.

(*c*) Berkley Co. Employee's Credit Union

1. Rate of interest paid: 5.50% simple interest paid on the *lowest* day's ending balance.
2. No service charge for checks processed.
3. No charge for check blanks.

(*d*) Washington Savings Association

1. Rate of interest paid: 5.50% compounded on the average daily ending balance.
2. Service charge of 10¢ on all checks processed.
3. All depositors must purchase check blanks. The cost is $7.00 for 100 checks (7¢ per check).

Calculate the net interest income earned or net cost that would be incurred by the Sammons for each of these checking account plans. Use a 30-day bank statement period in your calculations. Consider a year to have 365 days. Use the Interest Compounded Daily table on page 143 to calculate interest compounded on the average daily ending balance.

Solution: (*a*) Franklin National Bank

Step 1. Calculate the interest earned.

$$\text{Interest earned} = \text{principal} \times \begin{matrix}\text{interest compounded daily on} \\ \$1.00 \text{ at } 5.25\% \text{ for 30 days}\end{matrix}$$

$$\text{Interest earned} = \$600 \times .0043241$$

$$\text{Interest earned} = \$2.59$$

Step 2. Calculate the service charge. (None is assessed because the account's daily ending balance, $600, exceeds the bank's $500 minimum requirement.)

Step 3. Calculate the cost of buying check blanks.

$$\text{Cost of checks} = \begin{matrix}\text{number of} \\ \text{checks processed}\end{matrix} \times \begin{matrix}\text{cost} \\ \text{per check}\end{matrix}$$

$$\text{Cost of checks} = 40 \times 4¢$$

$$\text{Cost of checks} = 40 \times \$.04$$

$$\text{Cost of checks} = \$1.60$$

Step 4. Calculate the net interest income or net cost.

$$\begin{matrix}\text{Net interest} \\ \text{income or net cost}\end{matrix} = \begin{matrix}\text{interest} \\ \text{income earned}\end{matrix} - \left(\begin{matrix}\text{service} \\ \text{charge}\end{matrix} + \begin{matrix}\text{cost of} \\ \text{checks}\end{matrix}\right)$$

$$\begin{matrix}\text{Net interest} \\ \text{income}\end{matrix} = \$2.59 - (0 + \$1.60)$$

$$\begin{matrix}\text{Net interest} \\ \text{income}\end{matrix} = \$2.59 - \$1.60$$

$$\begin{matrix}\text{Net interest} \\ \text{income}\end{matrix} = \$.99$$

(*b*) Martin County Savings and Loan Association

Step 1. Calculate the interest earned.

$$\text{Interest earned} = \text{principal} \times \text{rate} \times \text{time}$$

$$\text{Interest earned} = \$600 \times 5.50\% \times \frac{30}{365}$$

$$\text{Interest earned} = \$600 \times .055 \times .0822$$

$$\text{Interest earned} = \$2.71$$

Step 2. Calculate the service charge.

$$\text{Service charge} = \text{number of checks processed} \times \text{charge per check}$$
$$\text{Service charge} = 40 \times 15¢$$
$$\text{Service charge} = 40 \times \$.15$$
$$\text{Service charge} = \$6.00$$

Step 3. Calculate the cost of buying check blanks.

No charge

Step 4. Calculate the net interest income or net cost.

$$\text{Net interest income or net cost} = \text{interest income earned} - \left(\text{service charge} + \text{cost of checks}\right)$$
$$\text{Net cost} = \$2.71 - (\$6.00 + 0)$$
$$\text{Net cost} = \$2.71 - \$6.00$$
$$\text{Net cost} = -\$3.29$$
$$\text{Net cost} = \$3.29$$

(*c*) Berkley Co. Employee's Credit Union

Step 1. Calculate the interest earned.

$$\text{Interest earned} = \text{principal} \times \text{rate} \times \text{time}$$
$$\text{Interest earned} = \$150 \times 5.50\% \times \frac{30}{365}$$
$$\text{Interest earned} = \$150 \times .055 \times .0822$$
$$\text{Interest earned} = \$.68$$

Step 2. Calculate the service charge.

No charge

Step 3. Calculate the cost of buying check blanks.

No charge

Step 4. Calculate the net interest income or net cost.

$$\text{Net interest income or net cost} = \text{interest income earned} - \left(\text{service charge} + \text{cost of checks}\right)$$
$$\text{Net interest income} = \$.68 - (0 + 0)$$
$$\text{Net interest income} = \$.68$$

(*d*) Washington Savings Association

Step 1. Calculate the interest earned.

$$\text{Interest earned} = \text{principal} \times \begin{array}{c}\text{interest compounded daily on}\\ \$1.00 \text{ at } 5.50\% \text{ for 30 days}\end{array}$$

$$\text{Interest earned} = \$600 \times .0045304$$

$$\text{Interest earned} = \$2.72$$

Step 2. Calculate the service charge.

$$\text{Service charge} = \text{number of checks processed} \times \text{charge per check}$$

$$\text{Service charge} = 40 \times 10¢$$

$$\text{Service charge} = 40 \times \$.10$$

$$\text{Service charge} = \$4.00$$

Step 3. Calculate the cost of buying check blanks.

$$\text{Cost of checks} = \text{number of checks processed} \times \text{cost per check}$$

$$\text{Cost of checks} = 40 \times 7¢$$

$$\text{Cost of checks} = 40 \times \$.07$$

$$\text{Cost of checks} = \$2.80$$

Step 4. Calculate the net interest income or net cost.

$$\begin{array}{c}\text{Net interest}\\ \text{income or net cost}\end{array} = \begin{array}{c}\text{interest}\\ \text{income earned}\end{array} - \left(\begin{array}{c}\text{service}\\ \text{charge}\end{array} + \begin{array}{c}\text{cost of}\\ \text{checks}\end{array}\right)$$

$$\text{Net cost} = \$2.72 - (\$4.00 + \$2.80)$$

$$\text{Net cost} = \$2.72 - \$6.80$$

$$\text{Net cost} = -\$4.08$$

$$\text{Net cost} = \$4.08$$

After the calculations are performed, the results should be listed and studied. In the foregoing example, the net interest income or net cost of checking services for a typical month for the Sammons would be as follows:

(*a*) Franklin National Bank	+$ .99
(*b*) Martin County Savings and Loan Association	+ .68
(*c*) Berkley Co. Employee's Credit Union	− 3.29
(*d*) Washington Savings Association	− 4.08

Now, with this type of information in hand, you will be able to make a logical choice of the interest-bearing checking account plan that is best for you. In the preceding illustration, it can be seen that the Sammons would be $5.07 ahead each month by selecting the plan offered by Franklin National Bank instead of that available from the Washington Savings Association. This may not sound like much, but in a year's time it amounts to $60.84 and over 10 years it adds up to a savings of $608.40.

In addition to the net interest income earned or the net cost of service charges and buying check blanks, other factors also should be considered before making a final selection. They include the convenience and proximity of the various financial institutions and the other services they offer, of which you may want to take advantage. Likewise, you should be aware that if your average daily ending balance, your lowest daily ending balance, or the number of checks you write changes substantially in the future from the "typical" information you have used in your analysis, a different interest-bearing checking account plan might then be best for you.

Calculating the *Real* Cost of an Overdraft Service Charge

Question 3 of this book's introduction states that a $10 overdraft service charge assessed by a bank for a $2 overdraft, when a deposit was made the day after the overdraft occurred, amounts to a 182,500% interest rate. Perhaps you're still wondering if that answer is correct. It is.

When your checking account becomes overdrawn and the bank pays the check anyway, the bank is, in essence, loaning you the amount of the overdraft until you make a deposit. The overdraft service charge actually amounts to charging you interest for the loan of that money. (Admittedly, though, the bank also has some processing expense in notifying you of the overdraft.)

Interest rates are always stated in terms of a year's time. Since a deposit is usually made very soon after an overdraft occurs, the time of the "loan" is often just a few days. Thus, the interest rate, stated as an annual percentage rate, is high.

The formula for calculating interest on a loan, interest = principal × rate × time, is the same as that described in the preceding section. When the rate of interest charged for an overdraft is calculated, the "interest" is the overdraft service fee and the "principal" is the amount of the overdraft. "Time" is a fraction with the number of days for which the overdraft existed as the numerator and the number of days in a year, 365, as the denominator. The "rate" is the annual percent of the overdraft charge, and it is the element of the formula that is being calculated.

To calculate the rate, the interest formula is restated as follows:

$$\text{Rate} = \frac{\text{interest}}{\text{principal} \times \text{time}}$$

EXAMPLE Austin King's checking account became overdrawn by $2. His bank paid the check and assessed him a $10 overdraft service charge. One day after the overdraft occurred, Austin made a deposit to his account, bringing its balance back to a positive amount. Calculate the annual percentage rate of interest which the overdraft service charge represents.

Solution:

$$\text{Rate} = \frac{\text{interest}}{\text{principal} \times \text{time}}$$

$$\text{Rate} = \frac{\$10.00}{\$2.00 \times \frac{1}{365}}$$

$$\text{Rate} = \frac{\$10.00}{\frac{\$2.00}{1} \times \frac{1}{365}}$$

$$\text{Rate} = \frac{\$10.00}{\frac{\$2.00}{365}}$$

$$\text{Rate} = \frac{\$10.00}{.00547945}$$

$$\text{Rate} = 1{,}825.00$$

$$\text{Rate} = 182{,}500\%$$

Remember, to express an amount as a percent, move the decimal point two places to the right, as shown above.

Since the rate of interest you must pay because of an overdraft is extremely high, it is important to manage your checking account so that overdrafts do not occur.

chapter 9

Investment Calculations

"And on Wall Street today the bulls outnumbered the bears as the Dow Jones Industrial Average surged ahead $10\frac{1}{2}$ points." It's difficult to go through a day without hearing a comment similar to that on some radio or television newscast. Terminology pertaining to stocks, bonds, mutual funds, commodities, and other investments may seem perplexing, but in reality it is easy to understand—once it has been explained to you.

The information in this section may not immediately transform you into an investment wizard or a market analyst, but it will provide you with the ability to understand various investment areas, read market quotations, and make the most frequently used investment calculations. As a starter, here's lesson one: "Bulls" are investors with a positive attitude; they expect the economy to show strength, and they anticipate that the market prices will increase. "Bears" are investors with a negative investment attitude; they expect poor business performance and economic decline, and they therefore anticipate a decrease in market prices.

STOCKS

The ownership of corporations is divided into units called *shares of stock*. There are two classes of stock: *common stock* and *preferred stock*.

Common stock is most widely owned, and it is the type most often mentioned in newscasts and conversations. Owners of preferred stock receive preferential treatment: they are the first to receive part of the company's earnings, and they receive payment before the common stockholders if the company should terminate operations.

Reading Stock Market Quotations

Each day, information pertaining to the trading of common and preferred stocks appears in most city newspapers and in financial publications such as *The Wall Street Journal.* The mere ability to read stock quotations will help you to understand a good deal about the stock market.

EXAMPLE The following stock market quotations appeared in today's newspaper. Interpret each element.

52 Week				Yld.	P-E	Sales				Net
High	Low	Stock	Div.	%	Ratio	100's	High	Low	Close	Chg.
$32\frac{1}{4}$	$24\frac{1}{8}$	ALVNCO	2.05	6.8	9	740	$30\frac{1}{2}$	$29\frac{7}{8}$	30	$+\frac{1}{2}$
60	$58\frac{1}{2}$	BLMRMF	pf4.95	8.4	...	112	$59\frac{1}{4}$	59	59	...

Solution:

1. The numbers under the headings "52 Week," "High," and "Low" show the highest and lowest prices at which this stock sold in the past 52 weeks. The numbers represent dollar amounts, to which they are converted by changing the fractions to their decimal equivalents and adding dollar signs. Therefore, the highest price at which the first stock sold in the past 52 weeks is \$32.25 ($32\frac{1}{4}$ = \$32.25); the lowest price in the past 52 weeks was \$24.125 ($24\frac{1}{8}$ = \$24.125). (If necessary, refer to page 8 for an explanation of how to convert fractions to decimals.)
2. The letters under the heading "Stock" are abbreviations of companies' names. ALVNCO stands for Alvin Company; BLMRMF stands for Balmer Manufacturing. The letters pf after BLMRMF indicate that

the stock is preferred. If the letters pf do not appear, the stock is common, as ALVNCO stock is.

3. The number under "Div." shows the stock's estimated yearly dividend based on the amount of quarterly or semiannual dividend paid so far this year. For Alvin Company, the estimated dividend for the year is $2.05.
4. The number under "Yld. %" shows the percent of return investors will earn if they purchase the stock at the day's closing price. The percent of return is determined from the amount of the annual dividend and the price shown under "Close." Thus, the return on investment in Alvin Company is 6.8% ($2.05 ÷ $30.00 × 100 = 6.8%).
5. The "P-E Ratio" shows the relation of the stock's selling *price* to the amount of company *earnings* per share. (Earnings per share are found in the company's financial statements.) Thus, the day's closing price of Alvin Company stock, $30.00, is 9 times the earnings per share over the past 12 months. Notice that a P-E ratio is not shown for preferred stock.
6. The number under "Sales 100's" shows the number of shares of stock sold during the day's trading, in hundreds. Thus, 74,000 shares of Alvin Company stock were traded on this day (740 × 100 = 74,000).
7. The number under "High" shows the highest price at which a share of stock sold during the day. The highest price per share of Alvin Company was $30.50.
8. The number under "Low" shows the lowest price at which a share of this stock sold during the day. The lowest price per share of Alvin Company was $29.875.
9. The number under "Close" shows the price at which the last sale of the day was made. The last sale of Alvin Company was at $30.00 per share.
10. The number under "Net Chg." shows the net change in selling price from yesterday's close to today's close. Thus, the last sale of Alvin Company today was $.50 higher than the last sale yesterday. The dots under "Net Chg." for Balmer Manufacturing indicates that there was no change in the two days' closing prices.

Buying and Selling Stock

Common stock and preferred stock of major companies are bought and sold through organized marketplaces called *stock exchanges*. The most dominant of these is the New York Stock Exchange, and the reference "Wall Street" pertains to trading that takes place there.

To buy or sell stock, contact a stockbroker in your community who is affiliated with a trader on the stock exchange floor. Your local broker will contact the trader, who will make the purchase or sale for you and relay the information back to your stockbroker. The entire process is performed electronically and usually takes from only a few seconds to a few minutes to complete. There is a commission charge both when you buy and when you sell.

The price paid when you buy stock is calculated by multiplying the number of shares you buy by the price per share. The broker's commission charge is added to that.

The amount you receive when you sell stock (the *net proceeds*) is calculated by multiplying the number of shares you sell by the price per share. The broker's commission charge is deducted from that.

An attractive aspect of investing in stocks, particularly common stocks, is that the price per share may increase. The difference between the selling price and the purchase price, if positive, is called a *capital gain*. You might also suffer a capital loss.

EXAMPLE Dan Delaney purchased 100 shares of CAMCO stock at $32\frac{1}{4}$. The stockbroker's commission charge was \$46.50. Later, Dan sold the 100 shares at $49\frac{1}{2}$ and the broker's commission charge was \$58.25. Calculate (*a*) the total cost of purchasing the stock, (*b*) the net proceeds when the stock was sold, and (*c*) the gain or loss on the stock.

Solution: (*a*) Calculate the total cost of purchasing the stock.

$$\text{Total cost} = \left(\begin{matrix}\text{price}\\ \text{per share}\end{matrix} \times \begin{matrix}\text{no. of shares}\\ \text{purchased}\end{matrix}\right) + \begin{matrix}\text{broker's}\\ \text{commission}\end{matrix}$$

$$\text{Total cost} = (32\tfrac{1}{4} \times 100) + \$46.50$$

$$\text{Total cost} = (\$32.25 \times 100) + \$46.50$$

$$\text{Total cost} = \$3{,}225.00 + \$46.50$$

$$\text{Total cost} = \$3{,}271.50$$

(*b*) Calculate the net proceeds from the sale.

$$\text{Net proceeds} = \left(\begin{matrix}\text{price}\\ \text{per share}\end{matrix} \times \begin{matrix}\text{no. of shares}\\ \text{sold}\end{matrix}\right) - \begin{matrix}\text{broker's}\\ \text{commission}\end{matrix}$$

$$\text{Net proceeds} = (49\tfrac{1}{2} \times 100) - \$58.25$$

$$\text{Net proceeds} = (\$49.50 \times 100) - \$58.25$$

$$\text{Net proceeds} = \$4{,}950.00 - \$58.25$$

$$\text{Net proceeds} = \$4{,}891.75$$

(*c*) Calculate the gain or loss.

$$\text{Gain or loss} = \text{net proceeds} - \text{total cost}$$

$$\text{Gain or loss} = \$4{,}891.75 - \$3{,}271.50$$

$$\text{Gain} = \$1{,}620.25$$

Calculating Dividends Earned

From an investor's viewpoint, dividends are income earned from owning stock.

Preferred Stock Dividends. If the stock is preferred, there is a stated dividend rate which is ordinarily a percent of the stock's par value (sometimes called stated value). The *par value* is the value assigned to each share of preferred stock by the issuing company, but it does not necessarily indicate the value of the stock or the price at which the stock will sell on the market.

The amount of annual dividend earned per share of preferred stock is calculated by multiplying the stated dividend rate by the stock's par value. If a quarterly dividend is paid, it is one-fourth of the annual dividend amount. To calculate the total dividends earned, multiply the number of shares of stock owned by the dividend earned per share.

EXAMPLE Marge Allen purchased 50 shares of 8% preferred stock in Commet Corporation for $77 per share. Par value of the stock is $100. Commet just declared a quarterly dividend on its preferred stock. Calculate the amount of quarterly dividend Marge will receive.

Solution:

Step 1. Calculate the annual dividend per share.

$$\text{Annual dividend per share} = \text{stock's par value} \times \text{stated annual dividend rate}$$

$$\text{Annual dividend per share} = \$100 \times 8\%$$

$$\text{Annual dividend per share} = \$100 \times .08$$

$$\text{Annual dividend per share} = \$8.00$$

Step 2. Calculate the quarterly dividend per share.

$$\text{Quarterly dividend per share} = \text{annual dividend per share} \div \text{no. of quarters in a year}$$

$$\text{Quarterly dividend per share} = \$8.00 \div 4$$

$$\text{Quarterly dividend per share} = \$2.00$$

Step 3. Calculate the total quarterly preferred stock dividend.

$$\text{Total quarterly dividend} = \text{quarterly dividend per share} \times \text{no. of shares owned}$$

$$\text{Total quarterly dividend} = \$2.00 \times 50$$

$$\text{Total quarterly dividend} = \$100.00$$

Common Stock Dividends. A common stock dividend is usually stated as a dollar and cents amount per share. The total dividend earned is calculated by multiplying the amount of dividend per share by the number of shares owned.

EXAMPLE John Elgersma owns 500 shares of Mandon Mining Company common stock on which a quarterly dividend of $.30 (30¢) per share has just been declared. Calculate the total quarterly dividend John will receive.

Solution:

$$\text{Total quarterly dividend} = \text{quarterly dividend per share} \times \text{no. of shares owned}$$

$$\text{Total quarterly dividend} = \$.30 \times 500$$

$$\text{Total quarterly dividend} = \$150.00$$

Neither preferred stock dividends nor common stock dividends are paid automatically; a dividend must be declared by the company's board of directors. Preferred stock dividends *must* be paid before any common stock dividends can be paid. Therefore, companies always pay preferred stock dividends if at all possible. Some companies never pay common stock dividends. Instead, they "plough back" all of the profits into expanding the company. In turn, this makes the company more valuable, and common stock prices should rise—at least in theory. Dividends are usually paid quarterly or semiannually.

Calculating Annual Yield

The term *annual yield* refers to the percent of return that stockholders earn on their investment. It is calculated by dividing the annual amount of dividend received by the investor's total cost of buying the stock.

EXAMPLE Alice Prashner owns 200 shares of Lewis Corporation common stock which she purchased at \$16 per share. The total commission charge was \$34. In the past year, quarterly dividends of \$.32, \$.27, \$.26, and \$.37 were received. Calculate Alice's annual yield on her investment.

Solution:

Step 1. Calculate the total cost of the stock.

$$\text{Total cost} = \left(\begin{matrix}\text{price}\\ \text{per share}\end{matrix} \times \begin{matrix}\text{no. of shares}\\ \text{purchased}\end{matrix}\right) + \begin{matrix}\text{broker's}\\ \text{commission}\end{matrix}$$

$$\text{Total cost} = (\$16 \times 200) + \$34$$

$$\text{Total cost} = \$3{,}200 + \$34$$

$$\text{Total cost} = \$3{,}234$$

Step 2. Calculate the year's total dividends.

$$\text{Year's total dividends} = \text{total quarterly dividends received} \times \text{no. of shares owned}$$

$$\text{Year's total dividends} = (\$.32 + \$.27 + \$.26 + \$.37) \times 200$$

$$\text{Year's total dividends} = \$1.22 \times 200$$

$$\text{Year's total dividends} = \$244.00$$

Step 3. Calculate the annual yield.

$$\text{Annual yield} = \text{year's total dividends} \div \text{stock's total cost}$$

$$\text{Annual yield} = \$244.00 \div \$3{,}234.00$$

$$\text{Annual yield} = .0754$$

$$\text{Annual yield} = 7.54\%$$

Over-the-Counter Stocks

Stocks of many smaller companies are traded through a network known as the *over-the counter* (OTC) market instead of being traded on organized stock exchanges. Various stockbrokers throughout the country act as dealers in OTC stocks. Dealers quote *bid* and *asked* prices, with the bid being the amount you would sell at and the asked being your purchase price. The difference between the two amounts is the dealer's profit for handling the transaction.

An order to buy or sell an OTC stock is placed through a local stockbroker, who handles the transaction for you. A commission is charged for both the purchase and sale of an OTC stock.

EXAMPLE The following quotations appeared in today's over-the-counter market section of the newspaper. Interpret each element of the quotation.

Stock & Div.		Sales 100's	Bid	Asked	Net Chg.
AllCo	.20	43	$16\frac{3}{4}$	$17\frac{1}{2}$	$+\frac{1}{8}$

Solution:

1. The letters under the heading "Stock & Div." are an abbreviation of the company's name. AllCo stands for Allen Corporation. The number following the company name, .20, is the stock's estimated annual dividend based on the amount of quarterly or semiannual dividend paid so far this year.
2. The number under the heading "Sales 100's" is the number of shares of this stock that sold during the day, in hundreds. On this day, 4,300 shares of AllCo stock sold (43 × 100 = 4,300).
3. The number under "Bid," $16\frac{3}{4}$, is the price for which you could have sold a share of AllCo stock, $16.75.
4. The number under "Asked," $17\frac{1}{2}$, is the price for which you could have purchased a share of AllCo stock, $17.50.
5. The number under "Net Chg.," $+\frac{1}{8}$, is the difference between the last sale made yesterday and the last sale made today. In this example, AllCo's stock sold for $.125 ($12\frac{1}{2}$ cents) more per share today than it did yesterday.

Other calculations for over-the-counter stocks are identical with those given in the preceding sections for stocks traded on an organized exchange.

Understanding the Dow Jones Industrial Average

The Dow Jones Industrial Average (DJI) is a widely quoted indicator of how the stock market performed during the day, week, or other time period. It consists of the common stock prices of the 30 companies listed below:

In calculating the number known as the Dow Jones Industrial Average, the prices at which these companies' stocks traded are added together. Then the total is divided by a number, which is currently 1.230.* Occasionally, this divisor is revised because of a change in a listed company's stock price caused by a stock split or by one company on the list being replaced by another. Therefore, the DJI provides a comparable indicator from year to year. The DJI is calculated many times during the day.

*Source: *The Wall Street Journal*

STOCKS INCLUDED IN DOW JONES INDUSTRIAL AVERAGE*

Allied Corp.	General Electric	Owens-Illinois
Aluminum Co.	General Foods	Procter & Gamble
American Brands	General Motors	Sears, Roebuck
American Can Co.	Goodyear	Standard Oil of Calif.
American Express	Inco	Texaco
AT&T	IBM	Union Carbide
Bethlehem Steel	Int'l Harvester	United Technologies
Du Pont	Int'l Paper	U.S. Steel
Eastman Kodak	Merck	Westinghouse Electric
Exxon	Minnesota M&M	Woolworth

*Source: *The Wall Street Journal*

An increase or decrease in the DJI of, say, 5 or 6 points or less (as from 950.21 to 956.30) is considered fairly modest and will evoke little analytical response. An increase or decrease of perhaps 10 to 12 points is fairly large but not uncommon. A change in the Dow of, say, 20 points or more (as from 1230.92 to 1205.86) is substantial and is an indication that some major news has greatly influenced the market.

The Dow Jones Industrial Average is an indicator of the market's performance in general. Over 1,700 stocks are traded on the New York Stock Exchange. Even on a day when the Dow is up substantially, say, 12 or 14 points, many stocks, perhaps as many as 400 or 500, will decrease in price while others stay the same or show a gain.

EXAMPLE This evening's newscast says that the Dow Jones Industrial Average closed down 26.19 points. Which of the following is the type of response you would expect most investors to display?

a. Ho hum—another dull day of trading.

b. That's interesting—but pretty normal—quite modest, actually.

c. Holy cow! Disaster must have struck somewhere, somehow.

Solution: No doubt, answer *c* would be a fairly normal reaction, since a change of 26.19 points is very substantial. Apparently there are some fears that the economic and business outlook is not good. It was a great day for the bears!

BONDS

Corporations and government units such as cities and counties often borrow money by issuing bonds to investors. The issuer pays semiannual or annual interest to the bondholder (investor) during the bond's lifetime and agrees to redeem (buy back) the bond at its *face value,* or par value, on the due date. Ordinarily, bonds have a $1,000 face value. They often have due dates 15, 20, 30, or more years from their date of issue.

Reading Bond Quotations

Often the original bondholder does not keep the bond until its due date. Bonds are traded through local stockbrokers or bond brokers who are affiliated with traders on organized exchanges, much in the same way stocks are traded. A bond's price usually fluctuates a good deal during its lifetime. That is mostly because of changing interest rates and other investment opportunities that become available to investors.

Many bond price quotations are available daily in city newspapers and financial publications.

EXAMPLE Today's quotations of two corporate bonds are shown below. Interpret each element of the quotation for the first one, BntCo.

Bonds	Cur. Yld.	Vol.	High	Low	Close	Net Chg.
BntCo $12\frac{1}{2}$s99	12	46	$101\frac{1}{4}$	$99\frac{3}{4}$	$100\frac{7}{8}$	$+\frac{1}{4}$
Ctrmfg $6\frac{1}{4}$s06	cv	15	$74\frac{1}{2}$	$74\frac{1}{4}$	$74\frac{1}{4}$	. . .

Solution:

1. Under the heading "Bonds" the issuing company's name is abbreviated. BntCo stands for Bennett Corporation.
2. The $12\frac{1}{2}$s following the company name is the rate of interest paid on the bond, $12\frac{1}{2}$%.
3. The number 99 stands for the bond's due date, 1999. (The number 06 for Ctrmfg stands for the year 2006.)
4. The 12 under "Cur. Yld." shows the current yield, which is the rate of return (in this case, 12%) that an investor will earn on a bond bought at the day's close, $100\frac{7}{8}$. The "cv" under "Cur. Yld." for Crtmfg means that the bond is *convertible;* that is, it can be converted into, or exchanged for, a certain number of shares of the issuing company's stock if the bondholder wishes.
5. The number under "Vol." shows the volume (number) of bonds traded on this day, 46.
6. The number under "High," $101\frac{1}{4}$, is the highest price at which this bond sold during the day. Since bonds ordinarily have a $1,000 face value, the quotation is multiplied by 10 to determine the dollar amount at which the bond sold. First the quotation is converted to its decimal equivalent ($101\frac{1}{4} = 101.25$). Then the quotation is multiplied by 10 and a dollar sign is added ($101.25 \times 10 = \$1,012.50$). (If necessary, turn to page 8 for a review of converting fractions to decimals.)
7. The number under "Low," $99\frac{3}{4}$, is the lowest price at which this bond sold during the day, $997.50 ($99\frac{3}{4} = 99.75$; $99.75 \times 10 = \$997.50$).
8. The number under "Close," $100\frac{7}{8}$, is the price at which the last sale of the day was made, or $1,008.75 ($100\frac{7}{8} = 100.875$; $100.875 \times 10 = \$1,008.75$).
9. The number under "Net Chg.," $+\frac{1}{4}$, shows the net change from the last sale made yesterday (the close) to the last sale made today. Today's close was $2.50 higher than yesterday's ($\frac{1}{4} = .25$; $.25 \times 10 = \$2.50$).

Calculating Interest Earned

The amount of annual interest earned on a bond is calculated by multiplying the bond's stated interest rate by the bond's face value without regard to the price an investor may have actually paid for the bond.

EXAMPLE Diane Olson paid $850.75 for a Poe Publishing Company bond ($1,000 face value), which has a $9\frac{3}{4}$ percent rate of interest. Calculate the amount of interest income Diane will receive annually.

Solution:

$$\text{Annual interest income} = \frac{\text{bond's}}{\text{face value}} \times \frac{\text{bond's stated}}{\text{interest rate}}$$

$$\text{Annual interest income} = \$1{,}000 \times 9\tfrac{3}{4}\%$$

$$\text{Annual interest income} = \$1{,}000 \times .0975$$

$$\text{Annual interest income} = \$97.50$$

An investor is not required to pay a federal income tax on the interest earned on *municipal bonds* (those issued by cities, counties, and so on). Therefore, an investor who is in a high income tax bracket might find that his or her after-tax income from an investment in lower-interest-bearing municipal bonds will be higher than from higher-interest-bearing corporate bonds.

Calculating Current Yield

The actual interest rate earned by an investor, based on the actual cost of the bond, is called the *current yield.* This is calculated by dividing the annual interest earned by the cost per bond. In calculating the bond's cost, the commission charge is added to the price of the bond. The commission is relatively low, usually being $5 to $7 per bond. If only a few bonds are purchased, however, there may be a minimum commission charge of, say, $30.

Since the current yield is an investor's annual rate of return on the investment, it is a meaningful calculation that can be used when comparing one investment's potential to another.

EXAMPLE Rick Friday purchased several $9\frac{1}{4}\%$ Lawrey Corporation bonds ($1,000 face value) at $77\frac{1}{2}$ ($775) each. In addition, there was a commission charge of $5 per bond. Thus, Rick's total cost per bond was $780.00. Calculate Rick's current yield per bond.

Solution:

Step 1. Calculate the annual interest earned per bond.

$$\text{Annual interest income} = \begin{array}{c}\text{bond's}\\ \text{face value}\end{array} \times \begin{array}{c}\text{bond's stated}\\ \text{interest rate}\end{array}$$

$$\text{Annual interest income} = \$1{,}000 \times 9\tfrac{1}{4}\%$$

$$\text{Annual interest income} = \$92.50$$

Step 2. Calculate the current yield.

$$\text{Current yield} = \text{annual interest income} \div \text{total cost per bond}$$

$$\text{Current yield} = \$92.50 \div \$780.00$$

$$\text{Current yield} = .1186$$

$$\text{Current yield} = 11.86\%$$

It is interesting to note that the current yield in this example, 11.86%, is substantially higher than the stated rate, 9.25%, because the bond was purchased for less than its $1,000 face value.

Calculating Yield to Maturity

As preceding examples have shown, bonds are often purchased at prices higher or lower than the $1,000 face value. If the purchase price is higher than the $1,000 face value, say, $1,020, the difference, $20, is called a *premium*. If the purchase price is lower than the $1,000 face value, say, $750, the difference, $250, is called a *discount*.

On their due date, bonds are redeemed (paid off) at their $1,000 face value. Therefore, if a bond was purchased at a premium, say, $1,050, and is held until the due date, its value will gradually decrease until it reaches the face value of $1,000 on the due date. Similarly, if a bond was purchased at a discount, say, $800, and is held until the due date, its value will gradually increase until it reaches the face value ($1,000) on the due date. This gradual decrease or increase in the bond's value because of purchase at a premium or a discount is called *amortization.*

When an investor intends to own the bond until its due date, the decrease in the bond's value because of buying it at a premium, or the increase because of buying it at a discount, must be considered when calculating the rate of return. The annual rate of return from the date a bond was purchased until the due date, considering both the interest earned and the premium or discount, is called the *yield to maturity.* To calculate the yield to maturity, follow this procedure:

1. Calculate the amount of premium or discount, which is the difference between the purchase price and the $1,000 face value.
2. Determine the annual amount of premium or discount amortization by dividing the amount of premium or discount by the number of years until the due date.
3. Use the result of your calculation in step 2 in one of the following ways:
   a. Deduct the annual premium from the annual interest earned
   b. Add the annual discount to the annual interest earned
4. Divide the result in step 3 by the average investment in the bond from the date of purchase until the bond's due date. The average investment is calculated by adding the cost of the bond (purchase price plus commission) to the proceeds that will be received when the bond is sold (face value less commission) and dividing the result by 2.
5. The calculation performed in step 4 will show the yield to maturity, which is the average annual rate of return if the bond is held until its due date.

EXAMPLE 1 Janice Webber purchased a $1,000 face value 12% bond at 108 and paid a $5 commission charge. Therefore, the bond cost her $1,085. The amount of annual interest earned on the bond is $120. The bond matures in exactly 4 years. When Janice sells the bond, it is anticipated that a $5 commission charge will be assessed, so she will receive $995 as proceeds. Calculate (*a*) the premium or discount, (*b*) the annual premium or discount amortization, (*c*) the average annual investment, and (*d*) the yield to maturity.

Solution: (*a*) Calculate the premium or discount.

Premium = purchase price − face value

Premium = $1,080.00 − $1,000.00

Premium = $80.00

(*b*) Calculate the annual premium amortization.

Annual premium amortization = premium ÷ years to maturity

Annual premium amortization = $80.00 ÷ 4

Annual premium amortization = $20.00

(*c*) Calculate the average annual investment.

$$\text{Average annual investment} = \frac{\text{bond's cost when purchased} + \text{bond's proceeds when sold}}{2}$$

$$\text{Average annual investment} = \frac{\$1{,}085.00 + \$995.00}{2}$$

$$\text{Average annual investment} = \frac{\$2{,}080.00}{2}$$

$$\text{Average annual investment} = \$1{,}040{,}00$$

(*d*) Calculate the yield to maturity.

$$\text{Yield to maturity} = \frac{\text{Annual interest earned} - \text{Annual premium amortization}}{\text{average annual investment}}$$

$$\text{Yield to maturity} = \frac{\$120.00 - \$20.00}{\$1{,}040.00}$$

$$\text{Yield to maturity} = \frac{\$\ 100.00}{\$1{,}040.00}$$

Yield to maturity = .0962

Yield to maturity = 9.62%

EXAMPLE 2 Neil Woodstock purchased a $1,000 face value 8% bond at 75 and paid a $5 commission charge. Therefore, his bond cost him $755. The annual interest earned on the bond is $80. The bond matures in exactly 5 years, When Neil sells the bond, it is anticipated that a $5 commission charge will be assessed, so he will receive $995 as proceeds. Calculate (*a*) the premium or discount, (*b*) the annual premium or discount amortization, (*c*) the average annual investment, and (*d*) the yield to maturity.

Solution: (*a*) Calculate the premium or discount.

Discount = face value − purchase price

Discount = $1,000 − $750

Discount = $250

(*b*) Calculate the annual discount amortization.

Annual discount amortization = discount ÷ years to maturity

Annual discount amortization = $250 ÷ 5

Annual discount amortization = $50

(*c*) Calculate the average annual investment.

$$\text{Average annual investment} = \frac{\text{bond's cost when purchased} + \text{bond's proceeds when sold}}{2}$$

$$\text{Average annual investment} = \frac{\$755 + \$995}{2}$$

$$\text{Average annual investment} = \frac{\$1{,}750}{2}$$

Average annual investment = $875

(*d*) Calculate the yield to maturity.

$$\text{Yield to maturity} = \frac{\text{Annual interest earned} + \text{Annual premium amortization}}{\text{average annual investment}}$$

$$\text{Yield to maturity} = \frac{\$80 + \$50}{\$875}$$

$$\text{Yield to maturity} = \frac{\$130}{\$875}$$

$$\text{Yield to maturity} = .1486$$

$$\text{Yield to maturity} = 14.86\%$$

It can be seen from these examples that the yield to maturity is often a far different, and more meaningful, rate of return than is indicated by the stated interest rate or current yield.

MUTUAL FUNDS

We are all familiar with the fact that companies such as General Motors and IBM earn their profits by manufacturing and selling products. A *mutual fund,* on the other hand, is an investment company in which investors pool their money in the hope of earning a profit by buying and selling stocks and bonds of companies such as General Motors and IBM and other investment mediums as well. Most mutual funds own securities of at least 20 different companies, and many mutual funds invest in well over 100 different companies.

Most investment companies offer several different mutual funds, each of which is designed to fulfill a specific investment purpose. A conservative investor might select a mutual fund that trades only in bonds, only in preferred stocks, or in both. An investor who is seeking long-term growth might choose a mutual fund that trades in common stocks of large, well-established companies (blue-chip stocks). An aggressive investor can select a mutual fund that trades in common stocks of more progressive companies and thus hope to realize substantial growth in a shorter time. Whatever your personality and investment objectives, you will be able to find many different mutual fund offerings suited to your tastes. A mutual fund's prospectus (a report) is a statement of the fund's purpose. It lists the companies in which securities are currently owned.

Mutual funds that trade in bonds and preferred stocks earn much of their profit from interest and dividend income. They can also earn a profit by selling the bonds or preferred stocks at prices higher than those paid, although bonds and preferred stocks ordinarily do not increase in value rapidly. Mutual funds that trade in common stocks make much of their profit by actively buying and selling stocks of various companies. They also earn dividends on many of the stocks owned.

In effect, when you buy a share of a mutual fund, you acquire, although indirectly, small pieces of ownership in the securities of the companies in which the mutual fund has invested. Diversification of your investment dollars among many different companies is one of the prime advantages of investing in a mutual fund.

Reading Mutual Fund Quotations

The value of a mutual fund share is called the *net asset value* (NAV). The mutual fund calculates it by starting with the total value of investments owned, deducting all liabilities (debts), and dividing the result by the total number of shares owned by its investors.

A *no-load mutual fund* does not charge a commission. A mutual fund which charges a commission (usually in the 7% to 9% range, but it can be higher) is called a *load fund.* The procedure for identifying whether a mutual fund is a load or a no-load fund is described in the following example and solution.

EXAMPLE The following mutual fund quotations appeared in today's newspaper. Interpret each element.

	NAV	Offer Price	NAV Chg.
ALLIED FUNDS GROUPS:			
Cont. Fund	10.18	13.14	+.08
Flag. Fund	27.06	N.L.	−.14

Solution:

1. The heading in capitals, ALLIED FUNDS GROUP, identifies the name of the investment company that manages each of the mutual funds listed below it.
2. The name "Cont. Fund" is the abbreviation for Continental Fund. The abbreviation "Flag. Fund" stands for Flagstaff Fund.
3. Since for Continental Fund an amount is shown under the "NAV" and also the "Offer Price" heading, this is a load fund which charges a sales commission. An investor would buy at the offer price, $13.14 per share. If an investor were to sell shares in Continental Fund, the sale would be at the NAV price, $10.18 per share. The difference between the two amounts is Continental Fund's commission charge.

   Flagstaff Fund is a no-load fund, as indicated by the letters "N.L." under the offer price heading. The NAV $27.06 is the amount a buyer would pay and a seller would receive.
4. The amount shown under the "NAV Chg." heading is the change in the mutual fund's net asset value from yesterday's closing quotation to today's closing quotation. Today's closing quotation for Continental Fund is $.08 (8¢) higher than yesterday's. Flagstaff Fund closed $.14 (14¢) lower than yesterday.

Buying and Selling Mutual Funds

Investment companies offer various plans whereby an investor can buy shares of a mutual fund. One method is to make a lump sum purchase of a certain number of shares or to invest a certain dollar amount. In another program, called *dollar cost averaging,* the investor agrees to invest a certain dollar amount each month for an extended time such as 8, 10, or 12 years. The investor can, however, discontinue the plan at any time.

EXAMPLE 1 Ellen Van Dorsten invests $50 per month, through a dollar-cost-averaging program, in Canton Mutual Fund, which is a no-load fund. On the 25th of this month, Ellen's usual purchase date, the NAV of Canton Fund was $10.85. Calculate the number of shares of Canton Fund Ellen purchased this month. Round your answer off to three decimal places.

Solution:

$$\text{Number of shares purchased} = \text{amount invested} \div \text{purchase price per share}$$

$$\text{Number of shares purchased} = \$50.00 \div \$10.85$$

$$\text{Number of shares purchased} = 4.608$$

EXAMPLE 2 Peter Ramos purchased 125 shares of LeMar Mutual Fund 4 years ago, when the NAV was $18.56 and the offer price was $20.25. Today the NAV is $26.50 and the offer price is $28.91. Calculate (*a*) the total cost when the shares were purchased, (*b*) the proceeds if the shares were sold today, and (*c*) the profit or loss that would result if the shares were sold today.

Solution: (*a*) Calculate the total cost.

Since this is a load fund, the purchase was made at the offer price.

$$\text{Total cost} = \text{purchase price per share} \times \text{number of shares purchased}$$

$$\text{Total cost} = \$20.25 \times 125$$

$$\text{Total cost} = \$2{,}531.25$$

(*b*) Calculate the proceeds; the sale would be made at the NAV.

$$\text{Proceeds} = \text{sales price per share} \times \text{number of shares sold}$$

$$\text{Proceeds} = \$26.50 \times 125$$

$$\text{Proceeds} = \$3{,}312.50$$

(*c*) Calculate the gain or loss on the sale.

$$\text{Gain or loss} = \text{proceeds} - \text{total cost}$$

$$\text{Gain} = \$3{,}312.50 - \$2{,}531.25$$

$$\text{Gain} = \$781.25$$

The value of mutual fund shares increases or decreases with the value of the companies' securities in which the mutual fund has invested. Thus, if you have invested in a common stock mutual fund and the general trend of the stock market is up, chances are your mutual fund shares also will show a gain.

BANK ACCOUNTS

Passbook Savings Accounts

Passbook savings accounts offer the investment advantages of being convenient and simple to understand. There is no minimum balance that must be maintained, and the investor can make deposits or withdrawals at any time without penalty. The investor's "cost" of receiving those advantages is that a lower interest rate is earned than is available from most other investments.

On most passbook savings accounts, interest is compounded on each day's ending balance. Typically, the rate is 5.00%, 5.25%, or 5.50%. Financial institutions use computers to calculate the interest, but an accurate answer can be reached by using a compound interest table like the one given here. Table 9-1 shows the amount of interest earned on an investment of $1.00 at various interest rates and for various time periods.

To calculate the amount of interest earned on your passbook savings account, proceed as follows:

1. Read down the "Days" column until you come to the number of days for which the money is invested.
2. Read across to the right until you reach the column with an interest rate heading that is the same as the interest earned on your account. The number in that column is the amount of

TABLE 9-1 INTEREST EARNED ON $1.00 WHEN COMPOUNDED DAILY

	Stated interest rate			
Days	5.00%	5.25%	5.50%	5.75%
1	$.0001370	$.0001438	$.0001507	$.0001575
30	.0041178	.0043241	.0045304	.0047368
31	.0042553	.0044685	.0046818	.0048951
90	.0124022	.0130284	.0136530	.0142779
365	.0512675	.0538986	.0565362	.0591805

interest earned on $1.00 at that interest rate for that number of days.

3. Multiply the number selected from the table in step 3 by the amount invested in your account. The answer is the amount of interest earned.

EXAMPLE Kim Burte invested $2,000 in a passbook savings account that pays 5.25% interest compounded daily. How much interest will she earn in a 30-day month?

Solution:

Step 1. Find the interest on $1.00 at 5.25% for 30 days. From Table 9-1 we see that the amount is $.0043241.

Step 2. Calculate the interest earned.

Interest earned	=	amount invested	×	interest on $1.00 at 5.25% for 30 days
Interest earned	=	$2,000	×	.0043241
Interest earned	=		$8.65	

It is interesting to calculate the amount of interest earned at various interest rates. For instance, if in the preceding example the rate had been 5.50% compounded daily, the interest income would have been $9.06 for the month.

Money Market Investments

Most financial institutions offer popular investment programs called *money market investments.* One of these is the *money market certificate,* whereby the investor must commit to investing a minimum amount for a minimum time period. A popular money market certificate calls for a $2,500 minimum investment, with available time periods ranging from 6 months to 3 years. Increasingly higher interest rates are offered for longer-term commitments.

Simple interest is paid on money market certificates, and new interest rates are set each week. Once an investor purchases a money market certificate, however, the rate of interest in effect at the time of the purchase applies for that certificate's duration. If the investor withdraws the funds before the end of the agreed term, a penalty of several months' interest is assessed.

Table 9-2 shows how money market certificate interest rates vary with the length of time for which an investment commitment is made. The typical penalty assessed for early withdrawal of the funds also is shown.

The interest rates listed in Table 9-2 are only examples, and you would have to check with local financial institutions to determine current rates. Many financial institutions also offer a money market certificate with a \$500 minimum investment and a 12-month time period. The interest rate on this plan is lower (perhaps $\frac{1}{2}$% lower) than the rate paid on a 6-month certificate for \$2,500.

The simple interest formula, interest = principal × rate × time (I = PRT), is used to calculate the amount of interest earned on a money market certificate, and it is also used to calculate any penalty for early withdrawal.

TABLE 9-2 MONEY MARKET CERTIFICATES*

Length of investment period	Simple interest rate earned	Penalty for early withdrawal
6 months to 364 days	10.00%	3 months' interest
1 year to 1 year, 364 days	10.50%	6 months' interest
2 years to 2 years, 364 days	10.75%	6 months' interest

*2,500 minimum investment

EXAMPLE Barry Wells invested \$4,000 in a 6-month money market certificate that pays 10.15% interest. A penalty of 3 months' interest is assessed if the certificate is redeemed before the maturity date. Calculate (*a*) the proceeds if the certificate is held to maturity and (*b*) the proceeds if the certificate is redeemed after 2 months.

Solution: (*a*) Calculate the proceeds if the certificate is held to maturity (the full 6 months).

Step 1. Calculate the interest earned.

$$\text{Interest} = \text{principal} \times \text{rate} \times \text{time}$$

$$\text{Interest} = \$4{,}000 \times 10.15\% \times \frac{6}{12}$$

$$\text{Interest} = \frac{\$4{,}000 \times .1015 \times 6}{12}$$

$$\text{Interest} = \frac{\$2{,}436}{12}$$

$$\text{Interest} = \$203$$

Step 2. Calculate the proceeds.

$$\text{Proceeds} = \text{principal} + \text{interest}$$

$$\text{Proceeds} = \$4{,}000 + \$203$$

$$\text{Proceeds} = \$4{,}203$$

(*b*) Calculate the proceeds if the certificate is redeemed after 2 months.

Step 1. Calculate the interest earned until the certificate is redeemed (2 months interest).

$$\text{Interest} = \text{principal} \times \text{rate} \times \text{time}$$

$$\text{Interest} = \$4{,}000 \times 10.15\% \times \frac{2}{12}$$

$$\text{Interest} = \frac{\$4{,}000 \times .1015 \times 2}{12}$$

$$\text{Interest} = \frac{\$812}{12}$$

$$\text{Interest} = \$67.67$$

Step 2. Calculate the certificate's value when redeemed.

$$\text{Value when redeemed} = \text{principal} + \text{interest}$$

$$\text{Value when redeemed} = \$4{,}000 + \$67.67$$

$$\text{Value when redeemed} = \$4{,}067.67$$

Step 3. Calculate the penalty for early redemption.

$$\text{Penalty} = \text{principal} \times \text{rate} \times \text{time}$$

$$\text{Penalty} = \$4{,}000 \times 10.15\% \times \frac{3}{12}$$

$$\text{Penalty} = \frac{\$4{,}000 \times .1015 \times 3}{12}$$

$$\text{Interest} = \frac{\$1{,}218}{12}$$

$$\text{Interest} = \$101.50$$

Step 4. Calculate the proceeds.

$$\text{Proceeds} = \text{certificate's value when redeemed} - \text{penalty}$$

$$\text{Proceeds} = \$4{,}067.67 - \$101.50$$

$$\text{Proceeds} = \$3{,}966.17$$

It can be seen from part (*b*) of the example that early redemption of a money market certificate may result in the proceeds being less than the original investment. Therefore, you should not invest in a money market certificate, or a similar investment with a stated time period, unless you are fairly certain that you will not need the funds during the investment period.

Another type of money market investment is the *high-interest, federally insured account* (sometimes referred to as Hi-Fi), which amounts to a passbook account that earns money market rates. Usually, a minimum of $2,500 must be maintained in the account at all times to receive the money market interest rate. The rate is somewhat lower than that paid on money market certificates, but it is considerably higher than that paid on regular passbook savings accounts.

The investor can deposit funds in or withdraw funds from a Hi-Fi account without penalty. If the balance drops below the $2,500 minimum, the account is treated as a regular passbook savings account and the passbook savings account interest rate applies.

COMMODITIES

Various commodities, such as wheat, lumber, orange juice, cocoa, soybeans, corn, sugar, gold, and silver are traded on the futures market. Purchases and sales of these commodities are made in terms of *contracts* of a certain number of units. The following are examples of the units contained in various commodities contracts: wheat, 5,000 bushels; orange juice, 15,000 pounds; cocoa, 10 metric tons; cotton, 50,000 pounds; lumber, 130,000 board feet; and silver, 5,000 troy ounces.

Futures market contracts call for delivery of the commodities to the buyer on some specific date in the future, which might be several months or more than a year from the purchase date. Persons who use the commodities market as an investment medium, however, do not actually intend to take delivery of the commodities, and they therefore buy back the contracts before the delivery dates. If the price per bushel, pound, ounce, or other unit of the contract has increased between the purchase date and the date on which the contract is bought back, the investor makes a profit. If the price has decreased, the investor suffers a loss.

Those who invest in commodities contracts are more properly called speculators than investors, since their goal ordinarily is to earn a large profit in a short time period. The risk of a large loss, however, is also a real possibility.

Reading Futures Market Quotations

Futures market quotations are available daily in financial publications and many newspapers. The quotations show trading that has occurred in commodities markets such as the Chicago Mercantile Exchange, Chicago Board of Trade, and Commodity Exchange, New York.

EXAMPLE The futures market quotations below appeared in a January issue of a newspaper. Interpret each element.

WHEAT (CBT) 5,000 bu.; Cents per bu.

	Open	High	Low	Settle	Change
Mar.	385	386	382	$382\frac{1}{4}$	-4
May	$395\frac{1}{2}$	$396\frac{1}{4}$	$392\frac{1}{2}$	$393\frac{1}{2}$	-3
July	$395\frac{1}{2}$	$396\frac{3}{4}$	393	$393\frac{1}{2}$	$-3\frac{3}{4}$
Sept.	405	405	400	400	$+\frac{1}{2}$
Dec.	419	419	419	419	$+3$

Solution:

1. The heading indicates that the quotations that follow are for wheat, that there are 5,000 bushels per contract, and that the numbers shown for prices indicate cents per bushel. The letters CBT mean that this commodity is traded on the Chicago Board of Trade.
2. Each month listed in the leftmost column represents a separate wheat contract and identifies the month in which delivery of the commodity will be made.
3. The number under the heading "Open" is the price per bushel, in cents, of the first sale of the day. Thus for a March contract, the price was 385 cents, or $3.85, per bushel.
4. The number under the heading "High" is the highest price per bushel at which trading occurred (for example, $3.86 for March).
5. The number under the heading "Low" is the lowest price per bushel at which trading occurred (for example, $3.82 for March).
6. The number under the heading "Settle" is the price per bushel at which the last sale of the day was made (for example, $\$3.82\frac{1}{4}$, or $3.8225, for March).
7. The number under the heading "Change" is the change in price per bushel from yesterday's settle price to today's settle price (for example, 4 cents lower for March).

Buying and Selling Futures Contracts

The value of a futures contract is calculated by multiplying the number of units in a contract by the price per unit.

One of the great appeals of speculating in the futures market is that commodities can be purchased on *margin*. That is, the buyer makes only a small down payment, say, 10% of the contract's value. Therefore, the speculator's actual cash investment is low compared with the total value of the investment controlled.

A commission charge in the range of $25 to $35 is paid to a broker both when a commodities contract is bought and when it is sold.

To calculate the margin requirement, multiply the contract's value by the margin requirement percent. To determine the amount of cash needed to buy a futures contract, add the margin requirement to the commission charge.

To calculate the gain or loss when the contract is sold, calculate the value of the contract at the time of sale and deduct the value of the contract at the time of purchase. Include the broker's commission in your calculations.

EXAMPLE Kermit Hauber purchased a 5,000-bushel contract of wheat at $4.00 per bushel. The margin requirement was 10%, and the broker's commission was $30. Two days later, when the price of wheat was $4.20 per bushel, Kermit sold the contract. The sales commission charge was $30. Calculate (*a*) the total value of the contract purchased, (*b*) the dollar amount needed for the margin requirement, (*c*) the cash needed to buy the contract, (*d*) the total value of the contract when sold, and (*e*) the gain or loss in this 2-day investment.

Solution: (*a*) Calculate the contract's value when purchased.

$$\text{Contract's value when purchased} = \begin{array}{c}\text{no. of units}\\ \text{per contract,}\\ \text{bushels}\end{array} \times \begin{array}{c}\text{price per}\\ \text{bushel}\end{array}$$

$$\text{Contract's value when purchased} = 5{,}000 \times \$4.00$$

$$\text{Contract's value when purchased} = \$20{,}000$$

(*b*) Calculate the margin requirement.

$$\text{Margin requirement} = \begin{array}{c}\text{contract's value}\\ \text{when purchased}\end{array} \times \begin{array}{c}\text{margin}\\ \text{percent}\end{array}$$

$$\text{Margin requirement} = \$20{,}000 \times 10\%$$

$$\text{Margin requirement} = \$20{,}000 \times .10$$
$$\text{Margin requirement} = \$2{,}000$$

(*c*) Calculate the cash needed to buy the contract.

$$\text{Cash needed} = \text{margin requirement} + \text{commission charge}$$
$$\text{Cash needed} = \$2{,}000 + \$30$$
$$\text{Cash needed} = \$2{,}030$$

(*d*) Calculate the contract's value when sold.

$$\text{Contract's value when sold} = \begin{matrix}\text{no. of units}\\ \text{per contract,}\\ \text{bushels}\end{matrix} \times \begin{matrix}\text{price per}\\ \text{bushel}\end{matrix}$$
$$\text{Contract's value when sold} = 5{,}000 \times \$4.20$$
$$\text{Contract's value when sold} = \$21{,}000$$

(*e*) Calculate the gain or loss on the investment.

$$\begin{matrix}\text{Gain}\\ \text{or loss}\end{matrix} = \begin{matrix}\text{contract's value}\\ \text{when sold}\end{matrix} - \left(\begin{matrix}\text{contract's value}\\ \text{when purchased}\end{matrix} + \begin{matrix}\text{total commission}\\ \text{charges}\end{matrix}\right)$$
$$\text{Gain} = \$21{,}000 - (\$20{,}000 + \$60)$$
$$\text{Gain} = \$21{,}000 - \$20{,}060$$
$$\text{Gain} = \$940$$

The example shows a highly successful investment whereby a cash outlay of $2,030 earned a $940 profit in just 2 days. That is a 7,459% gain stated as an annual percentage rate! It should be kept in mind, however, that if the price had decreased $.20 per bushel, there would have been a $940 loss (also at a 7,459% APR).

When the price, and therefore the value of the contract, drops, the investor will be faced with two alternatives neither of which is pleasant. One is to sell at a loss. The other, if the investor doesn't sell out, hoping for a price increase before the contract's delivery date, is to receive a *margin call.* That is, the investor will have to deposit more cash in his or her margin account to maintain the contract's original value.

Here is some sound advice: only an expert should consider speculating in the futures market.

INDIVIDUAL RETIREMENT ACCOUNTS

Most likely you've seen all of the advertisements by banks, stockbrokers, and others hawking the great investment opportunities they have available for your IRA. Do you know what that means—IRA? If not, it's something you should immediately find out, because an IRA could be the sweetest investment and income tax advantage you'll ever see.

The abbreviation IRA stands for the individual retirement account in which the Internal Revenue Service (IRS) allows an individual who has *earned income* to invest up to $2,000 per year. "Earned income" means income for which one's personal labors are responsible, such as wages earned as an employee or income earned by self-employment. Income that does not count includes interest, royalties, and dividends, all of which are called *unearned income.*

Again, any person who has taxable earned income of at least $2,000 a year may invest in an IRA. Therefore, if a husband and wife each has earned income of at least $2,000, each may fully invest in an IRA for a family total of $4,000. In many cases, one spouse, say, the husband, will have earnings in excess of $2,000 and the other spouse will have earnings of less than $2,000, say, $1,200. In that case, the husband can invest $2,000 in his IRA and the wife can invest up to the total of her earnings, $1,200, in her IRA. If one spouse has sufficient income to qualify but the other spouse has no income at all, a special *spousal IRA* may be utilized whereby a family total of $2,250 can be invested.

So what? you might be thinking. Well, the advantage of an IRA is that the wage earner is allowed by the IRS to subtract the amount invested in an IRA from taxable income for that year. Therefore, if a husband and wife have taxable income of $50,000 and file a joint federal income tax return, they can show taxable income of only $46,000 for that year if they are able to invest $4,000 in their IRAs. Therefore, their income tax will be calculated on earnings of $46,000 instead of $50,000, and they will pay less income tax in that year.

When a person takes advantage of the tax break afforded by an

IRA, the tax on that income is not forgiven. Instead, it is deferred, or postponed, until later. An individual must leave an IRA untouched until age $59\frac{1}{2}$. If the IRA is liquidated before that time, income tax must be paid on the proceeds in the year the IRA is liquidated. The tax rate that applies is that of the current year, and a substantial penalty (currently 10%) must be paid as well.

If the IRA is left untouched until after the person is $59\frac{1}{2}$, it can be liquidated without penalty. The investor will, however, have to pay federal income tax on the proceeds during the year the IRA is liquidated. The theory behind the IRA is that the person who does liquidate the investment will be at retirement age and his or her income will be far lower than it was during the years when the money was invested in the IRA. Therefore, the federal income tax rate on that money will be considerably lower and there will be an overall income tax saving. In addition, of course, the investor in an IRA has the advantage of saving a substantial amount of money for the retirement years.

To calculate the amount of federal income tax saving you can realize by investing in an IRA, multiply the amount invested by the highest tax rate that applies to your earnings.

EXAMPLE Last year the Hyles had taxable earned income of $54,000: Dallas earned $28,000, and Betty earned $26,000. They plan to file a joint federal income tax return. The upper portion of their income is taxed at the 50% rate. Calculate how much they will save in federal income taxes for the year if they invest the $4,000 maximum ($2,000 each) in IRAs.

Solution:

$$\text{Federal income tax saved} = \begin{matrix}\text{amount invested}\\ \text{in IRA}\end{matrix} \times \begin{matrix}\text{highest}\\ \text{tax rate}\end{matrix}$$

$$\text{Federal income tax saved} = \$4{,}000 \times 50\%$$

$$\text{Federal income tax saved} = \$4{,}000 \times .50$$

$$\text{Federal income tax saved} = \$2{,}000$$

The tax advantages provided by an IRA are greater for persons whose income is high and who are therefore taxed at the higher rates.

This discussion is intended to present the basic principles of an IRA. Consult your tax accountant for a more thorough explanation and an analysis of the benefits an IRA can provide for you. The Keogh plan and tax-sheltered annuities also provide income tax advantages similar to those of an IRA for persons in certain circumstances. Ask your tax accountant about them.

chapter 10

Borrowing Money

Nearly everyone finds it necessary to borrow money at one time or another, particularly when making a major purchase such as an automobile or a home. The information presented in this chapter should be particularly helpful in three ways: determining the length of time for which to obtain a loan, analyzing the effect of varying interest rates, and deciding whether to pay off part of the loan balance before it becomes due if you have extra cash available.

Determining the Monthly Payment

Most loans, particularly those for large amounts, are repaid in monthly installments. The amount of a monthly payment is determined by the amount borrowed, the interest rate charged, and the length of time for which the loan is obtained.

The loan repayment table (Table 10-1) shows the monthly payment required to repay loans for various amounts at 14% and 16% interest. If you are obtaining a loan for a different amount, different interest rate, or different repayment period, you can develop a very close estimate by using the amounts in the table as a guide. An example later in this chapter will show you how to read the loan repayment table.

TABLE 10-1 LOAN REPAYMENT TABLE*

Loan amount	14% interest rate 5 years	14% interest rate 20 years	14% interest rate 25 years	16% interest rate 5 years	16% interest rate 20 years	16% interest rate 25 years
$ 100	$ 2.33	$ 1.25	$ 1.21	$ 2.44	$ 1.40	$ 1.36
500	11.64	6.22	6.02	12.16	6.96	6.80
1,000	23.27	12.44	12.04	24.32	13.92	13.59
5,000	116.35	62.18	60.19	121.60	69.57	67.95
10,000	232.69	124.36	120.38	243.19	139.13	135.89
15,000	349.03	186.53	180.57	364.78	208.69	203.84
20,000	465.37	248.71	240.76	486.37	278.26	271.78
25,000	581.71	310.89	300.95	607.96	347.82	339.73
30,000	698.05	373.06	361.13	729.55	417.38	407.67
35,000	814.39	435.24	421.32	851.14	486.94	475.62
40,000	930.74	497.41	481.51	972.73	556.51	543.56
45,000	1,047.08	559.59	541.70	1,094.32	626.07	611.51
50,000	1,163.42	621.77	601.89	1,215.91	695.63	679.45
55,000	1,279.76	683.94	662.07	1,337.50	765.20	747.39
60,000	1,396.10	746.12	722.26	1,459.09	834.76	815.34
65,000	1,512.44	808.29	782.45	1,580.68	904.32	883.28
70,000	1,628.78	870.47	842.64	1,702.27	973.88	951.23
75,000	1,745.12	932.65	902.83	1,823.86	1,043.45	1,019.17
80,000	1,861.47	994.82	963.01	1,945.45	1,113.01	1,087.12
85,000	1,977.81	1,057.00	1,023.20	2,067.04	1,182.57	1,155.06
90,000	2,094.15	1,119.17	1,083.39	2,188.63	1,252.14	1,223.01
95,000	2,210.49	1,181.35	1,143.58	2,310.22	1,321.70	1,290.95
100,000	2,326.83	1,243.53	1,203.77	2,431.81	1,391.26	1,358.89

*Monthly payment required to pay off a loan

Determining the Total Repayment Amount

The first step in determining the total amount required to repay a loan over its lifetime is to calculate the total number of monthly payments that will be made. The number of monthly payments to be made is determined by multiplying the number of years in the repayment period by the number of months in a year, 12. This calculation is demonstrated in part (b) of the solution in the following section.

The total amount required to repay a loan is calculated by multiplying the number of payments to be made by the amount of each payment. This calculation also is demonstrated in part (b) of the solution in the folowing section.

Determining the Total Amount of Interest to Be Paid

The total amount of interest that will be paid on a loan is the difference between the total amount required to repay the loan and the amount borrowed.

EXAMPLE Henry Bruce plans to purchase a home for $75,000. He has $15,000 to use as down payment and will therefore borrow the remaining $60,000. A loan can be obtained at the 14% interest rate and can be repaid by making monthly payments for 25 years. Calculate (*a*) the monthly loan payment amount, (*b*) the total amount required to repay the loan over the 25 years, and (*c*) the total amount of interest that will be paid over the life of the loan.

Solution: (*a*) Calculate the amount of monthly payment.

Refer to the loan repayment table. Read down the "Loan amount" column until you come to $60,000. Then read across to the right until you come to the amount in the "25 years" column in the "14% interest rate" section. That amount, $722.26, is the monthly payment on the $60,000 loan at 14% interest for 25 years.

(*b*) Determine the total amount required to repay the loan.

Step 1. Calculate the total number of monthly payments to be made over the loan's lifetime.

$$\text{Total monthly payments} = \text{number of years for which loan is obtained} \times \text{number of payments in a year}$$

$$\text{Total monthly payments} = 25 \times 12$$

$$\text{Total monthly payments} = 300$$

Step 2. Determine the total repayment amount.

$$\text{Total repayment amount} = \text{monthly payment amount} \times \text{total monthly payments}$$

$$\text{Total repayment amount} = \$722.26 \times 300$$

$$\text{Total repayment amount} = \$216{,}678$$

(*c*) Calculate the total interest paid over the loan's lifetime.

$$\text{Total interest} = \text{total repayment amount} - \text{original amount of loan}$$

$$\text{Total interest} = \$216{,}678 - \$60{,}000$$

$$\text{Total interest} = \$156{,}678$$

Comparing Various Repayment Periods

Loans can often be obtained for various repayment periods. That is particularly true of long-term loans, such as home loans for which a 15-, 20-, or 25-year repayment period might be available. A comparison of the total amount to be repaid over the various time periods can be revealing, can help in your planning, and may well save you a great deal of money.

EXAMPLE A $50,000 home loan can be obtained at a 14% interest rate. Repayment can be made over either 20 years or 25 years. Calculate (*a*) the total repayment amount if the loan is obtained for 20 years, (*b*) the total repayment amount if the loan is obtained for 25 years, and (*c*) the total amount saved by selecting one repayment period instead of the other.

Solution: (*a*) Calculate the total repayment amount for a 20-year loan. (The monthly payments are $621.77. The total number of monthly payments is 240.)

$$\text{Total repayment amount} = \text{monthly payment amount} \times \text{total monthly payments}$$

$$\text{Total repayment amount} = \$621.77 \times 240$$

$$\text{Total repayment amount} = \$149{,}224.80$$

(*b*) Calculate the total repayment amount for a 25-year loan. (The monthly payments are $601.89. The total number of monthly payments is 300.)

$$\text{Total repayment amount} = \text{monthly payment amount} \times \text{total monthly payments}$$

$$\text{Total repayment amount} = \$601.89 \times 300$$

$$\text{Total repayment amount} = \$180{,}567.00$$

(*c*) Calculate the amount saved by selecting the 20-year loan repayment period.

$$\text{Amount saved} = \text{total repayment amount for 25-year loan} - \text{total repayment amount for 20-year loan}$$

$$\text{Amount saved} = \$180{,}567.00 - \$149{,}224.80$$

$$\text{Amount saved} = \$31{,}342.20$$

Comparing Various Interest Rates

On occasion, a lower interest rate can be obtained from one lender than from another. A lower interest rate will result in smaller monthly payments and a lower total amount to be repaid.

EXAMPLE Kelly Goodman plans to borrow $10,000 for the purchase of an automobile and will repay the loan over a 4-year period. The State Bank offers a 14% interest rate, and the monthly payments will be $273.27. Thompson Small Loan Company charges a 16% interest rate, and the monthly payments will be $283.41. Calculate (*a*) the total amount required to repay the 14% loan, (*b*) the total amount required to repay the 16% loan, and (*c*) the total amount saved by obtaining the 14% loan instead of the 16% loan.

Solution: (*a*) Determine the total repayment amount for the 14% loan.

$$\text{Total repayment amount} = \text{monthly payment amount} \times \text{number of payments}$$

$$\text{Total repayment amount} = \$273.27 \times 48$$

$$\text{Total repayment amount} = \$13{,}116.96$$

(*b*) Determine the total repayment amount for the 16% loan.

$$\text{Total repayment amount} = \begin{matrix}\text{monthly}\\ \text{payment amount}\end{matrix} \times \begin{matrix}\text{number}\\ \text{of payments}\end{matrix}$$

$$\text{Total repayment amount} = \$283.41 \times 48$$

$$\text{Total repayment amount} = \$13{,}603.68$$

(*c*) Determine the amount saved by obtaining the 14% loan.

$$\text{Amount saved} = \begin{matrix}\text{total repayment amount}\\ \text{for 16\% loan}\end{matrix} - \begin{matrix}\text{total repayment amount}\\ \text{for 14\% loan}\end{matrix}$$

$$\text{Amount saved} = \$13{,}603.68 - \$13{,}116.96$$

$$\text{Amount saved} = \$486.72$$

Calculating the Loan's Remaining Unpaid Balance

When a loan is repaid by making monthly payments over an extended period of time, earlier payments consist mostly of interest and very little of the loan balance is paid off. A little more of each monthly payment applies to reducing the loan balance because a little less interest is paid than in the preceding month.

The amount of the loan balance yet unpaid (the principal) can be calculated by referring to a *remaining loan balance table* like that shown as Table 10-2. The table shows the percent of the original loan that is still unpaid at the end of each year of the loan's term.

EXAMPLE Exactly 5 years ago, Marvel Epley purchased a home and borrowed $50,000 at 14% interest. The loan is to be retired on a 25-year repayment schedule. The monthly payments are $601.89. To date, Marvel has made 60 payments (5 years × 12 payments per year = 60). The total amount of her loan payments is $36,113.40 ($601.89 per month × 60 payments = $36,113.40). Calculate (*a*) the balance owed on the $50,000 loan, (*b*) the total amount of principal paid on the loan so far, and (*c*) the total amount of interest paid so far.

TABLE 10-2 REMAINING LOAN BALANCE TABLE FOR LOANS AT 14%

Age of loan	Original term of loan, in years								
	1	2	3	4	5	10	15	20	25
Year 1	0	53.47	71.18	79.95	85.15	95.06	97.89	99.02	99.53
2		0	38.07	56.91	68.08	89.38	95.46	97.89	98.98
3			0	30.43	48.46	82.85	92.67	96.59	98.35
4				0	25.91	75.35	89.46	95.09	97.63
5					0	66.73	85.77	93.38	96.80
6						56.82	81.53	91.40	95.85
7						45.43	76.66	89.13	94.76
8						32.34	71.06	86.53	93.50
9						17.29	64.63	83.53	92.05
10						0	57.23	80.09	90.39
11							48.73	76.13	88.48
12							38.97	71.58	86.28
13							27.74	66.36	83.76
14							14.83	60.35	80.86
15							0	53.44	77.53
16								45.51	73.70
17								36.38	69.30
18								25.90	64.23
19								13.85	58.42
20								0	51.73
21									44.05
22									35.22
23									25.07
24									13.41
25									0

Solution: (*a*) Calculate the remaining loan balance.

Step 1. Determine the percent of the loan balance still unpaid.

Refer to the remaining loan balance table. Read down the "25" column under the heading "Original term of loan, in years" until you come to the same line as 5 in the "Age of loan" column. The number, 96.80, is the percent of the original loan that is still unpaid.

Step 2. Determine the remaining loan balance.

$$\text{Remaining loan balance} = \begin{array}{c}\text{original amount}\\ \text{of loan}\end{array} \times \begin{array}{c}\text{percent of}\\ \text{loan unpaid}\end{array}$$

$$\text{Remaining loan balance} = \$50{,}000 \times 96.80\%$$

$$\text{Remaining loan balance} = \$50{,}000 \times .9680$$
$$\text{Remaining loan balance} = \$48{,}400$$

(*b*) Determine the amount of principal paid on the loan.

$$\text{Principal paid} = \begin{matrix}\text{original amount}\\ \text{of loan}\end{matrix} - \begin{matrix}\text{remaining}\\ \text{loan balance}\end{matrix}$$
$$\text{Principal paid} = \$50{,}000 - \$48{,}400$$
$$\text{Principal paid} = \$1{,}600$$

(*c*) Determine the total interest paid.

$$\text{Interest paid} = \text{total loan payments made} - \text{principal paid}$$
$$\text{Interest paid} = \$36{,}113.40 - \$1{,}600.00$$
$$\text{Interest paid} = \$34{,}513.40$$

Making Additional Principal Payments on the Loan

As can be seen by referring to the remaining loan balance table, the loan balance decreases very slowly in the early years of a long-term loan. In fact, on a 25-year loan, it takes about 20 years before half the loan is paid off. A great deal of progress is made in the last 5 years, however!

Since the loan balance decreases so slowly in the early years, it may be wise to make a payment directly on the loan's principal if you have additional cash available. That will, in essence, leapfrog you ahead on your loan payment schedule, will allow you to pay off the loan in less time than the loan's original repayment period, and will save you a great deal of money in interest payments.

EXAMPLE Exactly 4 years ago, Jay Wagner purchased a home and borrowed $60,000 at 14% interest. The loan was to be repaid on a 25-year schedule. Pertinent information about the loan, as calculated by using the procedures presented on the preceding pages, is as follows:

- Monthly loan payment: $722.26.
- Total amount required to repay the loan over 25 years: $216,678.
- Percent of loan balance unpaid at the end of 4 years: 97.63%
- Total amount of principal paid on loan at end of 4 years: $1,422.
- Amount of loan balance unpaid at the end of 4 years: $58,578.

At that point—after the loan had been in existence for 4 years—Jay made a $5,000 payment on the loan's remaining unpaid balance—that is, on the principal.

Calculate (*a*) the remaining loan balance after the $5,000 payment, (*b*) the point on the repayment schedule, in years, to which this payment advanced Jay, and (*c*) the total amount Jay saved by making the $5,000 payment.

Solution: (*a*) Calculate the remaining loan balance after the $5,000 payment.

$$\text{Remaining loan balance} = \text{previous loan balance} - \text{amount of advance payment on principal}$$

$$\text{Remaining loan balance} = \$58{,}578 - \$5{,}000$$

$$\text{Remaining loan balance} = \$53{,}578$$

(*b*) Determine the point on the repayment schedule, in years, to which the $5,000 payment advanced Jay.

Step 1. Calculate the percent of loan balance remaining unpaid after the $5,000 payment.

$$\text{Percent of loan balance remaining unpaid} = \text{remaining loan balance} \div \text{original amount borrowed}$$

$$\text{Percent of loan balance remaining unpaid} = \$53{,}578.00 \div \$60{,}000.00$$

$$\text{Percent of loan balance remaining unpaid} = .8930$$

$$\text{Percent of loan balance remaining unpaid} = 89.30\%$$

Step 2. Determine the age of the loan, in years, which corresponds with the percent of loan balance remaining unpaid.

Refer to the remaining loan balance table. Read down the 25 column under the "Original term of loan, in years" heading until you come to the number closest to the remaining loan balance percent calculated in step 1: 89.30. This number, 89.30, falls about in the middle of two amounts shown in the column: 90.39 and 88.48. Next read across to the left from those two numbers (90.39 and 88.48) until you come to the numbers in the "Age of loan" column. Those two numbers, 10 and 11, indicate the number of years on the loan repayment schedule to which Jay has advanced. Since the percent of unpaid principal on the loan, 89.30, falls between the two years, we can say that Jay has advanced to approximately $10\frac{1}{2}$ years on the loan repayment schedule.

Step 3. Determine the number of years of payments on the loan repayment schedule that have been eliminated by making the \$5,000 payment on the principal.

$$\text{Years of payments eliminated} = \text{current advancement on repayment schedule} - \text{previous advancement on repayment schedule}$$

$$\text{Years of payments eliminated} = 10\tfrac{1}{2} \text{ Years} - 4 \text{ Years}$$

$$\text{Years of payments eliminated} = 6\tfrac{1}{2} \text{ Years (or 78 months)}$$

(*c*) Calculate the total amount saved by making the \$5,000 payment.

Step 1. Calculate the total amount that would have been paid on the loan over the number of years that have been eliminated from the loan repayment schedule.

$$\text{Total amount that would have been paid} = \text{monthly loan payment} \times \text{number of months eliminated from loan repayment schedule}$$

$$\text{Total amount that would have been paid} = \$722.26 \times 78$$

$$\text{Total amount that would have been paid} = \$56{,}336.28$$

Step 2. Calculate the amount saved by making the \$5,000 payment.

$$\text{Amount saved} = \text{total amount that would have been paid} - \text{amount of this payment on principal}$$

$$\text{Amount saved} = \$56{,}336.28 - \$5{,}000.00$$

$$\text{Amount saved} = \$51{,}336.28$$

In summary, making a \$5,000 payment on the loan at the end of the fourth year has the effect of moving Jay ahead to approximately the $10\frac{1}{2}$-year point on the loan repayment schedule. That eliminates about $6\frac{1}{2}$ years of payments, and the loan will be paid off in full in 18½ years rather than the original 25 years. Making the $5,000 payment at the end of the fourth year will save more than $51,000 in payments over the life of the loan.

Usually, if the interest rate of a long-term loan is high, it is a good idea to consider making extra principal payments if you have the money to do so. Conversely, if the interest rate on your long-term loan is low, you may be able to put your money to better use elsewhere rather than make extra payments on your loan.

chapter 1

Preparing Personal Financial Statements

If you were to sell everything you own and pay off all of your debts, do you know how much cash would be left over for yourself? Here's another intriguing thought: A lot of dollars have probably passed through your hands during the past year; do you know where they all went? Perhaps you don't know the answers to those questions right offhand, but they can be easily obtained by completing two financial statements: the balance sheet and the income statement.

Satisfying your personal curiosity about your financial status is one good reason for preparing financial statements. Others are that it may be necessary to prepare the statements when you apply for a loan or when you develop a personal budget.

DEVELOPING A PERSONAL RECORD–KEEPING SYSTEM

The first step in preparing your personal financial statements is to accumulate accurate information about your income and expenditures. Having an organized system is the best way to do so, as described below.

Recording Income

Many people receive weekly or monthly paychecks with attached stubs which list current and year-to-date income and deductions for various taxes and for other reasons. The stubs can serve as a satisfactory record of your income, and no other system will be needed. Similarly, statements received from financial institutions can serve as your record of interest and dividend income earned. Simply accumulate these documents in an envelope or file folder for future use.

Those who have several sources of income or who desire a more sophisticated system, can record all income received on a form designed for the purpose, as outlined in the following steps:

1. Make a list of all sources of your income.
2. Purchase columnar accounting paper or draft a form with a sufficient number of columns to record all sources of income.
3. Label each column on your form with a title that describes a source of income (as shown in the example and solution).
4. Each time income is received:
   - *a.* Record the date.
   - *b.* Record the *total* amount earned (before deductions for taxes or voluntary payroll withholdings) in the "amount earned" column.
   - *c.* Record the total amount earned (same amount as shown in the "amount earned" column) under the appropriate income heading.
5. At the end of the month, year, or other period for which you are accumulating your data:
   - *a.* Total the "amount earned" column. This shows your total income from all sources.
   - *b.* Total each of the individual income amount columns.

c. To prove the accuracy of your work, add all of the individual income amount columns together. The total should equal the "amount earned" column total.

EXAMPLE John Mayer is a full-time teacher who receives two paychecks per month. He also works part time as a waiter, for which he receives tips and weekly wages. In addition, he has a savings account on which he earns interest. Income earned by John in September was as follows:

Sept. 6	Waiter's wages, $30; tips, $75 ($105 total)
13	Waiter's wages, $33; tips, $105 ($138 total)
15	Teacher's salary, $950.
20	Waiter's wages, $24; tips, $80 ($104 total)
27	Waiter's wages, $33; tips, $115 ($148 total)
30	Teacher's salary, $950
30	Interest income, $35

(*a*) Devise a form to record John's income. (*b*) Record John's income for the month. (*c*) Total the form at the end of the month. (*d*) Prove the accuracy of the column totals.

Solution: (*a*), (*b*), and (*c*) are shown in Table 11-1.

(*d*) Prove the column totals' accuracy.

Teacher's salary	$1,900	
Waiter's wages	120	
Tips	375	
Interest income	35	
	$2,430	(same as "amount earned" column total

Recording Expenditures

The type of form and the procedure for recording personal expenditures are very similar to those used for recording income, as described above. These are the steps to take:

1. Make a list of the various types of expenditure which you normally incur. Be sure to include everything for which you spend money. Most likely your list will contain many items—probably more than can be placed on one columnar form. Therefore, identify the 8 to 10 most frequently occurring expenditures that can be listed on a columnar form.
2. Draft a form to be used to record your expenditures. Allow sufficient space for the following. (Use the expenditures form in the example and solution as a guide.)
   a. Provide a column for the date on which the expenditure occurred.
   b. Provide three columns under an "amount paid" heading to provide for payments made by payroll deduction, by check, or by cash.
   c. Provide an individual column for each of your most frequently occurring expenditures.

TABLE 11-1

John Mayer Income For Month of September, 19—					
Date	Amount earned	Teacher's salary	Waiter's wages	Tips	Interest income
Sept. 6	$ 105		$ 30	$ 75	
13	138		33	105	
15	950	$ 950			
20	104		24	80	
27	148		33	115	
30	950	950			
30	35				35
Sept. Totals	$2,430	$1,900	$120	$375	$35

d. Provide a "miscellaneous" column in which to record the expenditures that are too inconsequential to identify by themselves.

e. Provide two columns, with the heading "other expenditures," in which to record expenditures that do not qualify for one of the individual columns or the miscellaneous column.

3. Purchase columnar accounting paper or draft a form to accommodate the number of columns required, as determined by the rough draft you developed in step 2.

4. Label the columns on your expenditures record.

5. Each time an expenditure occurs, do the following:

a. Record the date.

b. Record the amount of the expenditure in one of the "amount paid" columns. Use the "payroll deduction" column for items deducted by your employer, such as federal income tax, state income tax, FICA tax, and union dues. If payment was made by check, record the amount in the "check" column; if payment was made in cash, record the amount in the "cash" column.

c. If the expenditure was for an item for which you have provided an individual column, record the amount in that column.

If the expenditure is small and too inconsequential to classify by itself, record the amount in the "miscellaneous" column.

If the expenditure is for a substantial amount and/or should be identified by itself, but it does not fit into any of the individual columns, record it in the "other expenditures" column. To do so, write a one- or two-word description of the expenditure in the "description" column and put the amount in the "amount" column.

6. At the end of the month, year, or other period for which you are accumulating your data, follow this procedure:

a. Total each of the columns on your form.

b. To prove the accuracy of your work, add the totals of the three "amount paid" columns and then the totals of the remaining columns. The two totals should be equal.

EXAMPLE John Mayer's list of most frequently occurring expenditures is as follows: auto expense, groceries, utilities, clothing, and entertainment. The following are several expenditures which John incurred in September:

Sept. 28 Paid cash for groceries, $12.

28 Purchased a sport coat for $115; paid by check.

29 Donated $10 cash to the Boy Scouts.

30 Received paycheck from employer. The gross amount was $950. Deductions were as follows: federal income tax, $196; state income tax, $39; FICA taxes, $64. The total deductions are $299.

30 Purchased groceries for $38; paid by check.

(*a*) Devise a form for recording John's expenditures. (*b*) Record the five transactions listed above. (See Table 11-2 for the solution.)

After the expenditure record has been completed and the amount columns have been totaled, summarize the transactions in the "other expenditures" columns. That is, if two or more identical items are recorded, add the amounts together. Then list the summaries of the "other expenditures" columns on the back of your expenditure record for future use in preparing financial statements and for other purposes.

PREPARING A PERSONAL BALANCE SHEET

A *balance sheet* is a list of everything owned (called *assets*), all amounts owed (called *liabilities*), and the difference between the two (called *net worth*).

Calculating the Value of Assets

When you list assets on a personal balance sheet, show their *current net value:* the amount of cash the assets would bring if you were to

Solution:

TABLE 11-2

John Mayer Expenditure Record For Month Ending September 30, 19—											
	Amount Paid			Expenditures						Other Expenditures	
Date	Payroll Ded.	Check	Cash	Auto. Exp.	Groceries	Utilities	Clothing	Enter- tainment	Miscel- laneous	Description	Amount
Sept.											
28			12.00		12.00						
28		115.00					115.00				
29			10.00							Donations	10.00
30	299.00									Fed. Inc. Tax	196.00
										State Inc. Tax	39.00
										FICA Tax	64.00
30		38.00			38.00						

sell them, less any costs and expenses that would be incurred in making the sale.

The exact current net value of some assets, like savings accounts, checking accounts, the cash value of life insurance policies, and accounts receivable (amounts owed to you), can be determined by checking your own records or by inquiring of the appropriate institution or person.

The current net value of other assets, such as investments in stocks, bonds, and mutual funds, can be determined by checking current market quotations and asking a broker or agent for an estimate of sales expenses.

The exact current net value of many other assets will be difficult or impossible to determine, so you will need to estimate their value. For example, various price guide books might be used to estimate the current net value of antique vehicles, coins, stamps, and other collectibles.

The current net value of real estate can be estimated by having the property appraised by a real estate professional or by checking courthouse records for sale prices of comparable properties and then deducting the expected sales costs, as explained on pages 106 to 114.

The current net value of used household furnishings, furniture, appliances, and similar possessions can be difficult to estimate, since such property often decreases in value rapidly and may be difficult to sell readily. However, comparison with prices in second-hand stores or with prices obtained at auctions can help you make reasonable estimates.

The type and amount of assets owned will vary from one person to another, and your balance sheet should contain only the assets pertinent to your situation. The following list of personal assets will help you prepare your balance sheet, but you may have other items to include as well.

Commonly Owned Assets

Cash	Jewelry
Checking accounts	Household furniture
Savings accounts	Household furnishings

Commonly Owned Assets (Cont.)

Investments (stocks, bonds)	Appliances
Cash value of life insurance	Tools
Retirement and pension funds	Garden and lawn equipment
Accounts receivable	Vehicles
Deposits, which are refundable	Real estate
Antiques and collectibles	

Since it may be impossible to identify separately every item owned, the numerous items of small value each can be summarized in one category called "miscellaneous."

If there is an unpaid loan which originated from the purchase of an asset, do not deduct the loan balance from the value of the asset. Instead, list the loan separately in the "liabilities" section of the balance sheet.

Calculating Liabilities

To determine the balance owed on a charge account or credit card, refer to your most recent statement. Unpaid bills for medical care, utilities, taxes, and so on, also can be determined by referring to a recent statement.

If you have a loan, such as a bank loan, which calls for a lump sum repayment in the future, the amount owed can be determined by referring to your copy of the promissory note. You should also calculate the *accrued interest* on the loan and record that on the balance sheet. Accrued interest is the amount of interest expense that has accumulated on a loan from the date the loan originated to the date on which the balance sheet is prepared. Assume, for instance, that you obtained a loan to be repaid in 6 months and that $60 interest will be due when the loan is paid off. If 4 months have passed since you obtained the loan, $40 interest on the loan has accumulated ($\$60 \times \frac{4}{6} = \$40$), and you should record that $40 as accrued interest.

If you have an installment loan calling for monthly payments, you may have in your possession a schedule showing the loan balance remaining after each payment. Otherwise, for purposes of preparing a personal balance sheet, you can make a usable estimate of the remaining balance owed by multiplying the number of remaining payments by the amount per payment.

If you have a long-term real estate loan on which you make monthly or other periodic payments, you probably have a receipt showing the loan balance remaining after your last payment. Otherwise, you might have in your possession an amortization schedule which shows the remaining loan balance after each payment over the loan's lifetime. You can calculate the balance by following the procedure shown on pages 192 to 194, or you can ask the lender for an accurate figure.

The following list of liabilities will help you identify the liabilities to list on your balance sheet.

Possible Liabilities

Charge account balances
Credit card balances
Unpaid medical bills
Unpaid utility bills
Other unpaid bills and statements
Amounts owed to friends or relatives
Unpaid taxes
Bank loans
Installment loans
Real estate loans
Accrued interest

Calculating Net Worth

Net worth, which is also called capital and owner's equity, is the amount a person is worth financially. That is, if all assets were sold at their estimated current net values and all liabilities were paid, net worth is the amount of cash that would be left over. It is calculated by deducting the total liabilities from the total assets.

Total Liabilities and Net Worth

The final amount shown on a balance sheet is the *total liabilities and net worth.* It is calculated by adding the total liabilities to the net worth. The total should equal the total assets, and it is presented to show that the total claims to the assets (the liabilities and net worth) are equal to the total of the assets.

Preparing the Balance Sheet

Balance sheets prepared for businesses ordinarily have several subclasses for various types of assets and liabilities. On a personal balance sheet, however, it is permissible to simply list all assets under one heading and list all liabilities under another.

List cash as your first asset, followed by checking accounts, savings accounts, and other assets that could be easily turned into cash. Follow these with other assets loosely arranged by their length of life and ending with any real estate holdings. If you have a "miscellaneous" category, put it last.

List liabilities in the order in which the debts become due, beginning with those that become payable soonest and ending with the longest-term debts.

EXAMPLE Mr. and Mrs. Dan Carlson calculated the current net value of their assets on June 30 to be as follows: cash, $100; checking accounts, $500; savings accounts, $2,000; life insurance policy cash value, $2,400; jewelry, $800; automobiles, $7,000; household furniture, $6,000; household furnishings, $1,000; appliances, $2,500; real estate, $80,000, and miscellaneous, $500.

The Carlsons' debts are as follows: bank loan, $1,500; accrued interest on bank loan, $110; installment loan balance on auto, $5,000; real estate loan balance, $50,000.

Complete a balance sheet for the Carlsons.

Solution: The balance sheet prepared for the Carlsons would look like the one shown in Table 11-3.

A balance sheet shows the assets, liabilities, and net worth on a specific date. One prepared on even the following day would show somewhat different amounts because new transactions would have occurred.

PREPARING A PERSONAL INCOME STATEMENT

An *income statement* shows all income earned and all expenses incurred over some period of time such as a month or a year. Use

TABLE 11-3

MR. AND MRS. DAN CARLSON
BALANCE SHEET
JUNE 30, 19—

Assets		
Cash	$ 100	
Checking accounts	500	
Savings accounts	2,000	
Life insurance cash value	2,400	
Jewelry	800	
Household furnishings	1,000	
Household furniture	6,000	
Appliances	2,500	
Automobiles	7,000	
Real estate	80,000	
Miscellaneous	500	
Total assets		$102,800
Liabilities		
Bank loan	$ 1,500	
Accrued interest on bank loan	110	
Installment loan	5,000	
Real estate loan	50,000	
Total liabilities:		$ 56,610
Mr. and Mrs. Dan Carlson net worth		46,190
Total liabilities and net worth		$102,800

the *cash basis* when you prepare a personal income statement; that is, record only cash received as income and only cash paid out as expenses. Therefore, do not list any income earned but not yet received or any expenses incurred but not yet paid.

To prepare an income statement, you will need to keep accurate records of your income and expenses. Alternatively, you will need to search through your canceled checks, receipts, bills, statements, and other records to gather the information.

Calculating Income

When you are calculating personal income, include any method of earning money such as those listed below:

Sources of Income

Wages	Bonuses
Salary	Interest earned on savings accounts, bonds, and so on
Tips	Dividends earned on stocks, mutual funds, and so on
Commissions	

Record all income as the total amounts; deductions for income taxes, FICA tax, and so on will be recorded as expenses.

Calculating Expenses

In a personal income statement, expenditures for assets must be separated from expenditures for expenses. For example, payments made for the purchase of a vehicle, furniture, appliances, equipment, jewelry, investments, and other long-lasting items are not considered to be expenses; they are assets. On the other hand, payments made for items that are consumed or used up during the period are considered to be expenses. The following list will serve as a guide to the expenditures that should be listed as expenses. You may have other expense categories that should be included.

Possible Expenses

Federal income tax	Auto insurance
State income tax	Medical insurance
FICA tax	Groceries
Membership fees and dues	Entertainment
Rent	Clothing
Interest paid	Grooming and hygiene

Possible Expenses (Cont.)

- Electricity
- Water bill
- Sewer tax
- Natural gas
- Telephone
- Auto operation (gas, oil, repairs, etc.)
- Household repairs and maintenance
- Vacations
- Medical expenses
- Education
- Babysitting
- Homeowner's insurance

As a general rule, you should list separately only the expenses that are significant amounts by themselves or that you wish to identify separately. You might, for instance, list separately the electricity, water, sewer, and natural gas expenses, or you could lump them together under one heading, "utilities expense." Similarly, you might list auto insurance expense separately or include it in the auto operation category.

Undoubtedly, there will be many expenses that are too inconsequential to merit separate listing. They can be summarized in a single classification: miscellaneous expense.

Calculating Personal Net Income

Personal net income is calculated by deducting total expenses from total income. In most cases, the amount will not match the amount of cash you have on hand at the end of the period for which the income statement is prepared. That is because you have probably also spent money for the purchase of various assets like furniture, appliances, investments, or vehicles or have made payments on charge accounts and loans.

Preparing the Income Statement

An income statement prepared for a business might have numerous subdivisions for income and expenses. On a personal income statement, however, it is acceptable to list all income under one heading and all expenses under another.

If you have more than one source of income, list the sources in order by amount from largest to smallest. Similarly, list expenses in order by amount from largest to smallest or alphabetically with miscellaneous expense always last.

EXAMPLE Mary Diaz carefully recorded her income and expenses for the year ending December 31, 19—, which were as follows: Income: salary (gross), $26,000; interest income, $475. Expenses: federal income tax, $6,443; state income tax, $1,290; FICA tax, $1,742; rent, $4,800; utilities, $960; automobile, $980; groceries, $1,450; grooming and hygiene, $475; clothing, $1,960; medical, $170; entertainment, $1,225; and miscellaneous, $410. Use these data to prepare a personal income statement for Mary for the year.

Solution: The income statement would appear as shown in Table 11-4. The statement indicates that, after Mary pays all of her living expenses, $4,570 remains for saving, investing, purchasing assets, and repaying existing loans.

Alternate Personal Income Statement Form

Since an income statement is prepared largely for one's own use, the style can be varied to present the information in whatever you find to be the most usable and understandable format. One possibility, as briefly shown here, in Table 11-5, is to first deduct income and FICA taxes from total income to arrive at a "spendable income" amount. From that amount all other expenses are deducted to reach net income. The final result will be the same, regardless of the format used.

PREPARING A PERSONAL BUDGET

A *personal budget* is a guide to the amount of money a person expects to earn and the amounts that are to be spent for various purposes over a certain period of time.

TABLE 11-4

MARY DIAZ
INCOME STATEMENT
FOR YEAR ENDING DECEMBER 31, 19—

Income		
Salary	$26,000	
Interest	475	
Total income		$26,475
Expenses		
Federal income tax	$ 6,443	
Rent	4,800	
Clothing	1,960	
FICA tax	1,742	
Groceries	1,450	
State income tax	1,290	
Entertainment	1,225	
Automobile	980	
Utilities	960	
Grooming and hygiene	475	
Medical	170	
Miscellaneous	410	
Total expenses:		21,905
Net income		$ 4,570

Gathering Current Information

The starting point in preparing a budget is to gather accurate information on your past and current income and expenditures. Having an accurate recordkeeping system will be a big help here. If you do not have such a system, search your records and compile a list of all income and expenditures for the past year.

Record income from all sources including salary, wages, com-

TABLE 11-5

Income			
Salary	$26,000		
Interest	475		
Total income		$26,475	
Taxes			
Federal income tax	$ 6,443		
FICA tax	1,742		
State income tax	1,290		
Total taxes		−9,475	
Spendable income			$17,000
Expenses			
Rent	$ 4,800		
Groceries	1,450		

missions, tips, bonuses, interest earned, and dividends earned. Classify expenditures into specific categories such as federal, state, and FICA taxes, savings, investments, rent, house payments, real estate taxes, loan payments, charge account payments, food, clothing, and medical costs, household repair and maintenance, appliances, auto expenses, donations, and vacations. Lump all small expenditures together under "miscellaneous."

After you have a complete list of past and current income and expenditures, you can base reasonable projections for the future on those known amounts.

Determining Budget Amounts

Project your income for the forthcoming year by analyzing your current income from each source and modifying each amount by an expected pay raise or other change. Expenditures can be categorized as being either fixed or variable. *Fixed expenditures* are those that are set by either contract or circumstances and over which you have little or no control to change. They might include your mortgage payment, real estate taxes on your property, loan payment, and insurance premiums. *Variable expenditures* are those over which you have a good deal of control and which you can increase or decrease at your own discretion. Entertainment expenses, clothing purchases, and donations are examples of variable expenditures. Here are the steps to take to develop a budget for your expenditures:

1. List all of your fixed expenditures and the amount of each one.
2. Calculate a reasonable amount of your income for savings or investments.
3. Analyze each of your variable expenditures and predict a reasonable amount for each one.
4. Plan ahead. Identify all major expenditures which you plan to make during the coming year, such as the purchase of a new auto, and build those amounts into your budget.
5. Because of emergencies or unexpected events, it is impossible to predict accurately every expenditure that will be incurred during the coming year. Therefore, you may want to build a cushion into your budget and identify a certain sum for possible emergencies. If the events do not occur, you can invest the money or make a special purchase with it.
6. Because many expenditures that are not worth identifying by themselves may occur, budget a reasonable amount for "miscellaneous."

Preparing the Budget

One of several different formats may be used to present the budget. In one of them, shown in the example and solution, income and FICA taxes are deducted from total income to yield "spendable income." Then the total expenditure amount should equal the spendable income amount.

Ordinarily, it is a good idea to prepare a budget for a year's time and then break it down into a series of budgets for shorter periods of time such as a month. Some expenditures, such as an auto insurance premium, may occur only once or twice a year. Since it is necessary to plan during the entire year for such expenditures, it is wise to set aside portions of annual, semiannual, or quarterly expenditures in each month's budget.

A separate monthly account can be set up to accumulate the amounts needed for such expenditures. To break the expenditures down into monthly budgets, divide each amount by 12, 6, or 3, depending on whether payment is made annually, semiannually, or quarterly.

EXAMPLE Based on past and current information and future expectations, Stan and Betty McCord agreed on the following amounts for a personal budget for the coming year.

Income: Salary, $26,364; wages, $7,200; interest income, $36. Taxes: federal income taxes, $5,616; state income taxes, $1,128; FICA taxes, $2,256.

Expenditures: Savings, $1,200; mutual fund investment, $960; rent, $4,800; loan payments, $2,520; charge account payments, $540; food, $3,900; clothing, $1,500; grooming and hygiene, $480; utilities, $900; entertainment, $1,440; telephone, $480; auto expenses, $960; auto insurance premiums, $600; life insurance premiums, $480; household insurance premiums, $120; household supplies, $360; household furnishings, $1,020; household repair and maintenance, $300; donations, $480; vacations, $900; emergencies and unexpected expenditures, $420, and miscellaneous, $240. Prepare an itemized budget for the McCords, showing the amounts annually and monthly. See Table 11-6 for the solution.

Reviewing the Budget

After you have completed your budget, go back and review it. Ask yourself, "Are these amounts reasonable and logical?" If some amounts do not appear to be, change them before you make a final commitment to your budget.

TABLE 11-6

Stan and Betty McCord Personal Budget for January 1 to December 31, 19—				
Budget category	**Year**		**Month**	
Income				
Salary	$26,364		$2,197	
Wages	7,200		600	
Interest income	36		3	
Total income:		$33,600		$2,800
Less taxes				
Federal income taxes	5,616		468	
FICA taxes	2,256		188	
State income taxes	1,128		94	
Total taxes:		9,000		750
Spendable income:		$24,600		$2,050
Expenditures				
Rent	$ 4,800		$ 400	
Food	3,900		325	
Loan payments	2,520		210	
Clothing	1,500		125	
Entertainment	1,440		120	
Savings	1,200		100	
Household furnishings	1,020		85	
Mutual fund investment	960		80	
Auto expenses	960		80	
Utilities	900		75	
Vacations	900		75	
Auto insurance premiums	600		50	
Charge account payments	540		45	
Grooming and hygiene	480		40	
Telephone	480		40	
Life insurance premiums	480		40	
Donations	480		40	
Emergencies & unexpected exp.	420		35	
Household supplies	360		30	
Household repairs and maint.	300		25	
Household insurance premiums	120		10	
Miscellaneous	240		20	
Total expenditures:		$24,600		$2,050

Following the Budget

The purpose of a budget, remember, is to serve as a guide or goal for your income and expenditures. Therefore, you should make constant reference to your budget, particularly when you are contemplating expenditures, to determine whether you can afford them.

You should also keep accurate records of your actual income and expenditures and continually compare them with budgeted amounts. This will reveal whether or not you are following your budget.

Revising the Budget

A good budget should allow for a certain degree of flexibility so you can make adjustments as your circumstances, interests, and needs change. Assume, for example, that you have budgeted $150 per month for entertainment. Assume further that, sometime after your budget was prepared, you develop an interest in buying a pool table for your home. A logical way to afford the pool table might be to decrease your entertainment expenditure to $100 per month and use the other $50 budgeted for entertainment to make monthly payments on a loan secured to buy the pool table.

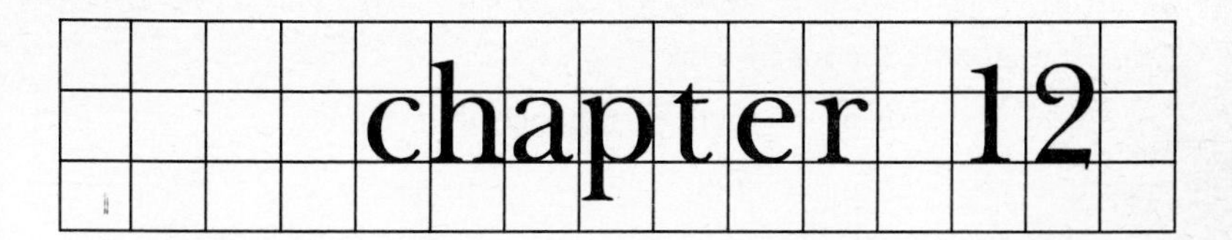

Calculating Statistics

Some people, particularly teachers, seem to lace their conversation with the statistical terms "mean," "median," and "mode." Each of these terms applies to a very basic calculation, one that you have probably done many times for your own use.

Calculating the mean, median, and mode, also called *measures of central tendency,* helps to show the relations of one amount to others in a series or group.

Calculating the Mean

The word "mean" has exactly the same definition as another word with which you are undoubtedly familiar: "average." To calculate the mean of a series of numbers, proceed as follows:

1. Add all of the amounts in the series.
2. Divide the total obtained in step 1 by the number of amounts in the series. The result is called the mean or average.

EXAMPLE Six students received the following scores on an examination: 96, 94, 88, 75, 70, and 66. Calculate the mean score.

Solution:

Step 1. Add the amounts.

$$96 + 94 + 88 + 75 + 70 + 66 = 489$$

Step 2. Calculate the mean score.

Mean = total ÷ number of amounts

Mean = 489 ÷ 6

Mean = 81.5

Calculating the Median

The *median* is simply the middle number in a series of numbers listed from largest to smallest or smallest to largest. Thus in the series 98, 90, 89 the median amount is 90. The median is useful when you want to know whether an amount falls in the top half or bottom half of a group.

The midpoint of a series can be identified by using this formula:

$$\text{Midpoint} = \frac{\text{number of amounts} + 1}{2}$$

If there is an even number of amounts in the series, the median can be stated as being an interval between the two numbers on either side of the midpoint. Another method is to show the median as an average of the two amounts.

EXAMPLE 1 Five students received the following scores on an examination: 90, 80, 86, 96, 88. Find the median.

Solution:

Step 1. List the amounts in order from highest to lowest.

96

90

88

86

80

Step 2. Determine the midpoint.

$$\text{Midpoint} = \frac{\text{number of amounts} + 1}{2}$$

$$\text{Midpoint} = \frac{5 + 1}{2}$$

$$\text{Midpoint} = \frac{6}{2}$$

$$\text{Midpoint} = 3$$

Step 3. Determine the median.

The median score is the third number on the list counting from highest to lowest or lowest to highest: 88

EXAMPLE 2 Four students received the following scores on an examination: 74, 94, 88, 84. Find the median.

Solution:

Step 1. List the amounts in order from highest to lowest.

94

88

84

74

Step 2. Determine the midpoint.

$$\text{Midpoint} = \frac{\text{number of amounts} + 1}{2}$$

$$\text{Midpoint} = \frac{4 + 1}{2}$$

$$\text{Midpoint} = \frac{5}{2}$$

$$\text{Midpoint} = 2.5$$

Step 3. Determine the median.

Since there is an even number of amounts in the series, the midpoint falls between two of the amounts listed. The midpoint calculated in step 2 is 2.5, which means that the median falls between the second and third amounts reading from highest to lowest or lowest to highest—that is, between 88 and 84. The median can be identified in one of two ways:

a. State the median as being an interval between the two amounts. That is, a median of 84 to 88.

b. Calculate an average of the two amounts.

$$\text{Median} = \frac{\text{total of amounts}}{2}$$

$$\text{Median} = \frac{88 + 84}{2}$$

$$\text{Median} = \frac{172}{2}$$

$$\text{Median} = 86$$

Calculating the Mode

The *mode,* or *modal score,* is simply the number that appears most often in the series. It shows the level of performance that occurs most frequently.

The mode can be determined by listing the amounts in order and making a tally of how often each number occurs.

EXAMPLE Students received the following scores on an examination: 94, 88, 92, 70, 88, 92, 88, 70, 90, 88, 96, and 92. Determine the mode.

Solution:

Step 1. List the amounts in order, making a tally.

96
94
92, 92, 92
90
88, 88, 88, 88
70, 70

Step 2. Identify the mode.

Since 88 occurs more frequently than any other amount, it is the mode.

How and Why Statistics Lie (Sometimes)

You have most likely heard this comment: "You can prove anything with statistics." Although that is not true, the fact is that the manner in which statistics are presented often can be very misleading. Witness these two examples:

EXAMPLE 1 While attending a class reunion, a man boasted to a former classmate, "Last year, the average earnings of my two brothers and myself was $160,000."

Solution: Technically, the statement was true, but it was grossly misleading. You see, one brother earned $200,000, another earned $260,000, and the boastful brother earned only $20,000. But, the average of the three brothers *was* $160,000!

EXAMPLE 2 From year A to year B, Shelby Company raised its base pay for employees from $9.00 to $9.05 per hour. From year B to year C, the base pay was raised from $9.05 to $9.14 per hour. Notice how differently this information appears on the graphs in Fig. 12-1(*a*) and (*b*) which contain exactly the same data.

Solution: Changing the wording of the heading and placing it in capital letters in the graph at the right makes the information appear more important. Showing amount increments by the cent, rather than the dollar, in the graph at the right makes the pay increases appear to be huge. You can probably spot at least one more technique that was used to make the graph at the right appear far more impressive than the graph at the left.

Fortunately, most people present data honestly and accurately. However, these examples demonstrate that additional information and study are sometimes necessary to interpret statistics accurately.

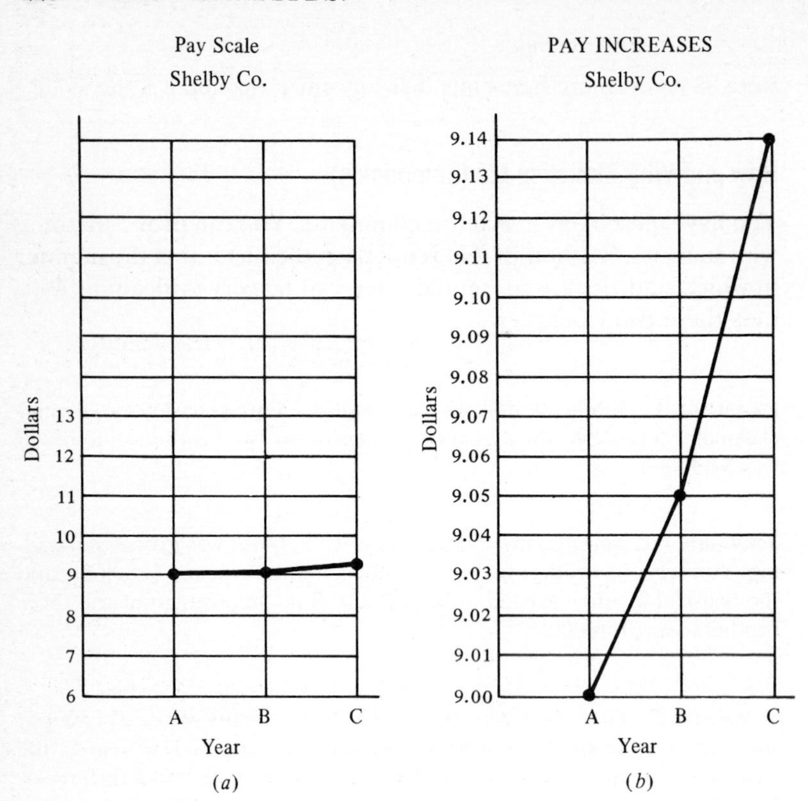
Pay Scale
Shelby Co.
Dollars
13
12
11
10
9
8
7
6
A
B
C
Year
(a)
PAY INCREASES
Shelby Co.
Dollars
9.14
9.13
9.12
9.11
9.10
9.09
9.08
9.07
9.06
9.05
9.04
9.03
9.02
9.01
9.00
A
B
C
Year
(b)

FIG. 12-1

chapter 13

Calculating Units of Measurement

The United States uses the English system of measurements. Most of the world, however, uses a form of the metric system known as the *International System of Units* (SI). Use of the metric system is becoming more prevalent in the United States, and most product labels now show measurements in both English and metric systems.

The purpose of this chapter is to identify commonly used English and metric measurements, show you how to convert from one unit to another within the same measurement system, and illustrate how to convert from one measurement system to the other.

ENGLISH MEASUREMENT CONVERSIONS

The English measurements listed in this section will serve as a handy reference, and the examples and solutions will be a good guide for making conversions from one unit to another.

Weight (Avoirdupois Measure)

Avoirdupois weight, which is 16 ounces to a pound, is used in measuring most goods. (Another weight measurement, *troy weight,* is 12 ounces to a pound. It is used to weigh precious metals and gems.)

TABLE 13-1 ENGLISH MEASUREMENT: AVOIRDUPOIS WEIGHT

Pennyweight	=	.05486 ounce
Ounce	=	18.2291 pennyweights
Ounce	=	.0625 pound ($\frac{1}{16}$ pound)
Pound	=	291.6667 pennyweights ($291\frac{2}{3}$)
Pound	=	16 ounces
Hundredweight	=	100 pounds
Ton (short)	=	2,000 pounds
Ton (short)	=	20 hundredweights
Ton (short)	=	32,000 ounces

EXAMPLE Convert the following weight measures to other units.

	Measurement	Convert to
(*a*)	20 ounces	pounds
(*b*)	1.75 pounds	ounces
(*c*)	3,500 pounds	short tons
(*d*)	4 short tons	pounds

Solution: Use Table 13-1 as a guide.

	Measurement	Converted to	Procedure
(*a*)	20 ounces	1.25 pounds	20 × .0625 (or, 20 ÷ 16)
(*b*)	1.75 pounds	28 ounces	1.75 × 16
(*c*)	3,500 pounds	1.75 short tons	3,500 ÷ 2,000
(*d*)	4 short tons	8,000 pounds	4 × 2,000

Liquid Measure

The commonly used measurements of liquids are pints, quarts, and gallons, although others listed in Table 13-2 are also used occasionally. In cooking, the teaspoon, tablespoon, and cup are common measurements.

TABLE 13-2 ENGLISH LIQUID MEASURE

Teaspoon	=	.16667 ounce ($\frac{1}{6}$ ounce)
Teaspoon	=	.3333 tablespoon ($\frac{1}{3}$ tbs.)
Tablespoon	=	3 teaspoons
Tablespoon	=	.5 ounce ($\frac{1}{2}$ ounce)
Tablespoon	=	.0625 cup ($\frac{1}{16}$ cup)
Cup	=	48 teaspoons
Cup	=	16 tablespoons
Cup	=	8 ounces
Cup	=	.5 pint ($\frac{1}{2}$ pint)
Cup	=	.25 quart ($\frac{1}{4}$ quart)
Gill	=	4 ounces
Gill	=	.25 pint ($\frac{1}{4}$ pint)
Gill	=	.125 quart ($\frac{1}{8}$ quart)
Gill	=	.03125 gallon ($\frac{1}{32}$ gallon)
Pint	=	16 ounces
Pint	=	4 gills
Pint	=	.5 quart ($\frac{1}{2}$ quart)
Pint	=	.125 gallon ($\frac{1}{8}$ gallon)
Quart	=	32 ounces
Quart	=	8 gills
Quart	=	2 pints
Quart	=	.25 gallon ($\frac{1}{4}$ gallon)
Gallon	=	128 ounces
Gallon	=	32 gills
Gallon	=	8 pints
Gallon	=	4 quarts
Barrel*	=	31.5 gallons

*A U.S. barrel of petroleum is 42 gallons.

EXAMPLE Convert the following liquid measurements to other units.

	Measurement	Convert to
(*a*)	9 teaspoons	tablespoons
(*b*)	.5 cup	tablespoons
(*c*)	48 ounces	pints
(*d*)	5 pints	quarts
(*e*)	3 quarts	pints

Solution: Use Table 13-2 as a guide.

	Measurement	Converted to	Procedure
(*a*)	9 teaspoons	3 tablespoons	9 × .333
(*b*)	.5 cup	8 tablespoons	.5 × 16
(*c*)	48 ounces	3 pints	48 ÷ 16
(*d*)	5 pints	2.5 quarts	5 × .5
(*e*)	3 quarts	6 pints	3 × 2

Dry Measure

Dry measure refers to goods other than liquids, such as grains, powders, and granulars. See Table 13-3.

TABLE 13-3 ENGLISH DRY MEASURE

Pint	=	.5 quart ($\frac{1}{2}$ quart)
Pint	=	.125 gallon ($\frac{1}{8}$ gallon)
Quart	=	2 pints
Quart	=	.25 gallon ($\frac{1}{4}$ gallon)
Quart	=	.125 peck ($\frac{1}{8}$ peck)
Quart	=	.03125 bushel ($\frac{1}{32}$ bushel)
Gallon	=	8 pints
Gallon	=	4 quarts
Gallon	=	.5 peck ($\frac{1}{2}$ peck)
Gallon	=	.125 bushel ($\frac{1}{8}$ bushel)
Peck	=	16 pints
Peck	=	8 quarts
Peck	=	2 gallons
Peck	=	.25 bushel ($\frac{1}{4}$ bushel)
Bushel	=	64 pints
Bushel	=	32 quarts
Bushel	=	8 gallons
Bushel	=	4 pecks

EXAMPLE Convert the following dry measurements to other units.

	Measurement	Convert to
(*a*)	36 gallons	bushels
(*b*)	8 quarts	pints
(*c*)	2 gallons	quarts
(*d*)	2 pecks	bushels
(*e*)	12 pints	gallons

Solution: Use Table 13-3 as a guide.

	Measurement	Converted to	Procedure
(*a*)	36 gallons	4.5 bushels	36 × .125
(*b*)	8 quarts	16 pints	8 × 2
(*c*)	2 gallons	8 quarts	2 × 4
(*d*)	2 pecks	.5 bushel	2 × .25
(*e*)	12 pints	1.5 gallons	12 × .125

Quantity Measure

The common measures of quantity are units, dozen, and gross. See Table 13-4.

TABLE 13-4 ENGLISH QUANTITY MEASURE

Unit	=	1
Unit	=	.08333 dozen ($\frac{1}{12}$ dozen)
Dozen	=	12 units
Dozen	=	.08333 gross ($\frac{1}{12}$ gross)
Gross	=	144 units
Gross	=	12 dozen

EXAMPLE Convert the following quantity measurements to other units.

	Measurement	Convert to
(*a*)	150 units	dozen
(*b*)	2 gross	dozen
(*c*)	18 dozen	gross
(*d*)	936 units	gross

Solution: Use Table 13-4 as a guide.

	Measurement	Converted to	Procedure
(*a*)	150 units	12.5 dozen	150 ÷ 12 (or 150 × .08333)
(*b*)	2 gross	24 dozen	2 × 12
(*c*)	18 dozen	1.5 gross	18 ÷ 12
(*d*)	936 units	6.5 gross	936 ÷ 144

Linear Measure (Length)

Measure of length is called *linear measure*. The inch, foot, and yard are the measures frequently used for common calculations. Other units of measurement, such as the rod and chain, are often used in land measurements. See Table 13-5.

EXAMPLE Convert the following linear measurements to other units.

	Measurement	Convert to
(*a*)	54 inches	feet
(*b*)	138 feet	yards
(*c*)	9,240 feet	miles
(*d*)	3.25 miles	yards
(*e*)	4 rods	feet
(*f*)	165 yards	rods
(*g*)	2 chains	feet

TABLE 13-5 ENGLISH LINEAR MEASURE

Link	=	7.92 inches
Foot	=	12 inches
Foot	=	.3333 yard ($\frac{1}{3}$ yard)
Yard	=	36 inches
Yard	=	3 feet
Rod	=	198 inches
Rod	=	16.5 feet ($16\frac{1}{2}$ feet)
Rod	=	5.5 yards ($5\frac{1}{2}$ yards)
Chain	=	100 links
Chain	=	66 feet
Chain	=	22 yards
Chain	=	4 rods
Chain	=	.10 furlong ($\frac{1}{10}$ furlong)
Furlong	=	660 feet
Furlong	=	220 yards
Furlong	=	40 rods
Furlong	=	10 chains
Furlong	=	.125 mile ($\frac{1}{8}$ mile)
Mile	=	5,280 feet
Mile	=	1,760 yards
Mile	=	320 rods
Mile	=	80 chains
Mile	=	8 furlongs

Solution: Use the linear measure conversion table as a guide.

	Measurement	Converted to	Procedure
(*a*)	54 inches	4.5 feet	54 ÷ 12
(*b*)	138 feet	46 yards	138 ÷ 3 (or 138 × .3333)
(*c*)	9,240 feet	1.75 miles	9,240 ÷ 5,280
(*d*)	3.25 miles	5,720 yards	3.25 × 1,760
(*e*)	4 rods	66 feet	4 × 16.5
(*f*)	165 yards	30 rods	165 ÷ 5.5
(*g*)	2 chains	132 feet	2 × 66

Circumference of a Circle. The distance around the outside of a circle is called the *circumference.* The circumference can be calculated by using the formula, $2 \pi R$. The symbol π, the Greek letter pi (pronounced like pie) is 3.1416. The R stands for *radius,* which is the distance from the center to the outer edge of a circle. (The radius is one-half of the circle's *diameter,* which is the distance across the circle through the circle's center.) Thus, the formula for calculating the circumference of a circle could be stated as $2 \times \pi \times R$, or $2 \times 3.1416 \times$ radius. (The circumference can also be calculated by multiplying pi by the circle's diameter D, as stated in the formula πD.)

EXAMPLE The radius of a circle is 14 inches. Determine the circle's circumference.

Solution:

$$\text{Circumference} = 2 \times \pi \times \text{radius}$$
$$\text{Circumference} = 2 \times 3.1416 \times 14$$
$$\text{Circumference} = 87.96 \text{ inches}$$

Square Measure

Square measure is used to find the number of square inches, feet, yards, or other units in a surface or area. See Table 13-6. If the area is a square or rectangle, the square measure is calculated by multiplying the area's length by its width. Thus, a square foot, which measures 12 inches, contains 144 square inches ($12 \times 12 = 144$). A parcel that measures 4 feet by 7 feet (a rectangle) contains 28 square feet ($4 \times 7 = 28$).

EXAMPLE Calculate the number of square units in the following areas. (The abbreviation L means length, and W means width.

	Measurement	Convert to
(*a*)	L = 8 feet; W = 6 feet	square feet
(*b*)	126 square feet	square yards
(*c*)	7 square yards	square feet
(*d*)	L = 24 inches; W = 18 inches	square feet

TABLE 13-6 ENGLISH SQUARE MEASURE

Square foot	=	144 square inches (12 inches × 12 inches)
Square foot	=	.11111 square yard (or $\frac{1}{9}$ sq. yd.) ($\frac{1}{3}$ yd. × $\frac{1}{3}$ yd.)
Square yard	=	1,296 square inches (36 inches × 36 inches)
Square yard	=	9 square feet (3 feet × 3 feet)
Square rod	=	272.25 square feet (16.5 feet × 16.5 feet)
Square rod	=	30.25 square yards (5.5 yards × 5.5 yards)
Square acre	=	43,560 square feet (208.7104 feet × 208.7014 feet)
Square acre	=	4,840 square yards (69.57 yards × 69.57 yards)
Square acre	=	160 square rods (12.65 rods × 12.65 rods)
Square mile	=	640 acres

Solution: Use Table 13-6 as a guide.

Measurement	Converted to	Procedure
(*a*) *L* = 8 feet; W = 6 feet	48 sq. ft.	8 × 6
(*b*) 126 square feet	14 sq. yds.	126 ÷ 9 (or, 126 × $.\overline{1}1111$)
(*c*) 7 square yards	63 sq. ft.	7 × 9
(*d*) L = 24 inches; W = 18 inches	3 sq. ft.	24 × 18 = 432 sq. in. 432 ÷ 144 = 3 sq. ft.

Square Measure of a Triangle. A three-sided area is called a *triangle.* If the vertical line of the triangle, the *height,* and the bottom horizontal line, called the *base,* form a right angle, the triangle is called a right-angled triangle, or simply a *right triangle.* To calculate the area of a right triangle, multiply the triangle's height by one-half its base, as expressed in the formula

$$\text{area} = \tfrac{1}{2}\,\text{base} \times \text{height}$$

EXAMPLE A parcel of land forms a right triangle with a height of 15 feet and a base of 20 feet as illustrated in Fig. 13-1. Calculate the number of square feet in the parcel.

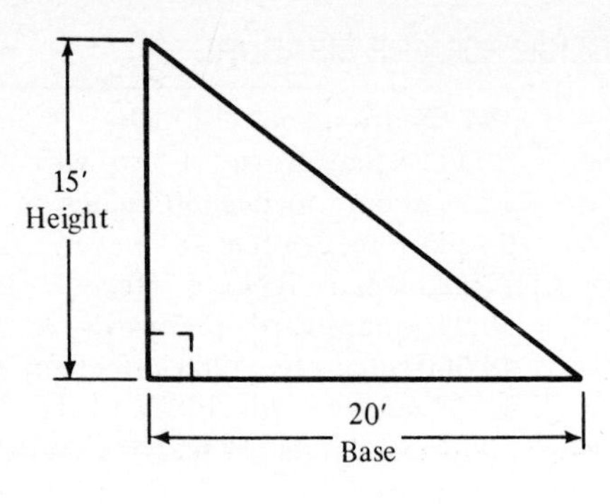

FIG. 13-1

Solution:

$$\text{Area} = \tfrac{1}{2}\text{ base} \times \text{height}$$
$$\text{Area} = \tfrac{1}{2}(20) \times 15$$
$$\text{Area} = 10 \times 15$$
$$\text{Area} = 150 \text{ square feet}$$

The area of a triangle that is not a right triangle can be calculated by drawing a line from the peak of the triangle straight down to the base so that it forms a right angle with the base. The result will be that two right triangles have been formed. Then, the area of each right triangle can be calculated by using the formula area = $\frac{1}{2}$ base × height. The area of the entire original triangle is determined by adding together the areas of the two right triangles.

EXAMPLE A parcel of land forms a triangle with a base to 40 feet as illustrated in Fig. 13-2. Calculate the number of square feet in the area.

Solution:

Step 1. Do the following:

a. Draw a line from the peak of the triangle to the base so that two right triangles are formed. (In Fig. 13-3 they are labeled A and B for ease of identification.)

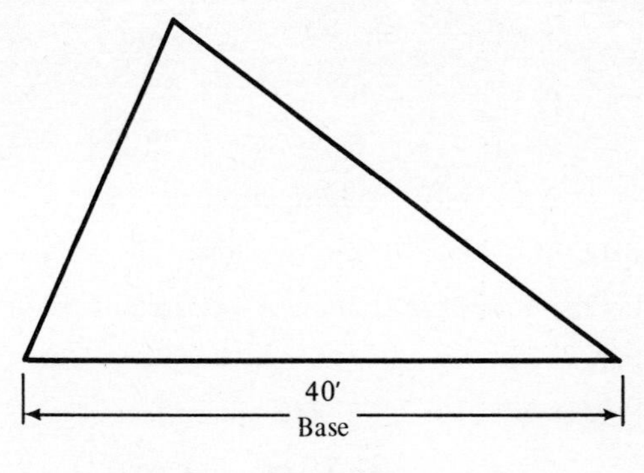

FIG. 13-2

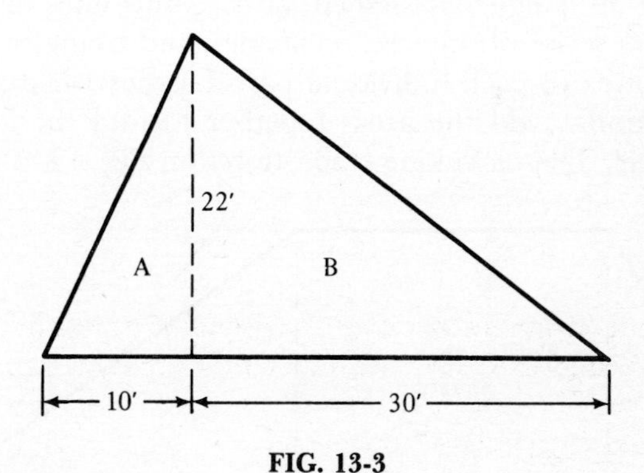

FIG. 13-3

b. Measure the height of the vertical line. (In this example, it is 22 feet.)

c. Measure the base of each new right triangle that has been formed (In this example, the base of A is 10 feet; the base of B is 30 feet; see Fig. 13-3.)

Step 2. Calculate the area of right triangle A.

$$\text{Area} = \tfrac{1}{2}\text{ base} \times \text{height}$$
$$\text{Area} = \tfrac{1}{2}(10) \times 22$$
$$\text{Area} = 5 \times 22$$
$$\text{Area} = 110 \text{ square feet}$$

Step 3. Calculate the area of right triangle B.

$$\text{Area} = \tfrac{1}{2}\text{ base} \times \text{height}$$
$$\text{Area} = \tfrac{1}{2}(30) \times 22$$
$$\text{Area} = 15 \times 22$$
$$\text{Area} = 330 \text{ square feet}$$

Step 4. Calculate the area of the entire parcel.

Total area = area of triangle A + area of triangle B

Total area = 110 + 330

Total area = 440 square feet

Square Measure of Irregularly Shaped Parcels. To calculate the area of an irregularly shaped parcel, draw lines to divide the parcel into a series of squares, rectangles, and triangles. Then calculate the area of each individual parcel as described in the preceding sections. Add the areas together to find the area of the entire parcel. The technique is illustrated in Fig. 13-4.

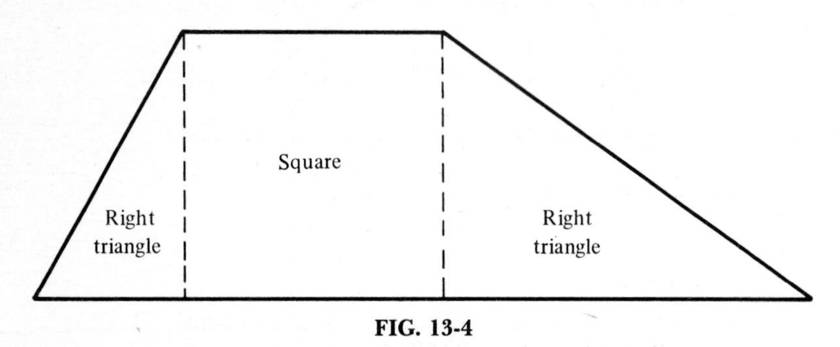

FIG. 13-4

Square Measure of a Circle. The square measure of a circle can be easily calculated with the formula πR^2 (read pi R square). Pi (π) is always 3.1416, and R stands for the circle's radius, which is the distance from the center to the outer edge of the circle. The term R^2 (radius square), means to multiply the radius by itself. Thus the formula could be stated as $\pi \times R \times R$, or $3.1416 \times R \times R$.

EXAMPLE Find the area of a circle which has a radius of 2 feet.

Solution:

$$\text{Area} = \pi \times R \times R$$

$$\text{Area} = 3.1416 \times 2 \times 2$$

$$\text{Area} = 12.5664 \text{ square feet}$$

Cubic Measure

Cubic measure is three-dimensional. An enclosed volume's length, width, and height are multiplied by each other to determine the number of cubic inches, feet, yards, or other units contained in the volume. Thus, a cubic foot measuring 12 inches in length, 12 inches in width, and 12 inches in height, contains 1,728 cubic inches ($12 \times 12 \times 12 = 1{,}728$). Cubic measure is particularly useful in determining the number of quarts, gallons, or other units a container can hold. See Table 13-7.

TABLE 13-7 ENGLISH CUBIC MEASURE

Cubic foot	=	1,728 cubic inches
Cubic foot	=	.037 cubic yard ($\frac{1}{27}$ cubic yard)
Cubic foot	=	25.7140 quarts (dry)
Cubic foot	=	29.922 quarts (liquid)
Cubic foot	=	6.4285 gallons (dry)
Cubic foot	=	7.4805 gallons (liquid)
Cubic yard	=	46,656 cubic inches
Cubic yard	=	27 cubic feet
Cubic yard	=	173.5699 gallons (dry)
Cubic yard	=	201.9740 gallons (liquid)
Cubic yard	=	29.6962 bushels

EXAMPLE Calculate the number of cubic units in the following areas, for which length (L), width (W), and height (H) are stated.

	Measurement	Convert to
(*a*)	L, 3 ft.; W, 4 ft.; H, 2 ft.	cubic feet
(*b*)	3 cubic yards	cubic feet
(*c*)	4 cubic feet	gallons (dry)
(*d*)	1.5 cubic yards	gallons (liquid)

Solution: Use Table 13-7 as a guide.

	Measurement	Converted to	Procedure
(*a*)	L, 3; W, 4; H, 2 (ft.)	24 cu. ft.	$3 \times 4 \times 2$
(*b*)	3 cubic yards	81 cu. ft.	3×27
(*c*)	4 cubic feet	25.714 gal. (dry)	4×6.4285
(*d*)	1.5 cubic yards	302.961 gal. (liq.)	1.5×201.9740

Cubic Capacity of a Cylinder The cubic capacity of a cylinder is also called the *volume* of a cylinder. The term *cylinder* refers to a round container such as a pail or barrel.

The formula to calculate the volume of a cylinder is $\pi R^2 H$. The symbol π, or pi, is 3.1416. R stands for the radius of the cylinder's circular top or botttom, and it is the distance from the center to the outer edge of the cylinder. The term R^2 (radius square) means to multiply the radius by itself. H stands for the height of the cylinder. Thus, the formula can be stated as: $\pi \times R \times R \times H$, or $3.1416 \times R \times R \times H$.

EXAMPLE The top of a barrel has a radius of 1.25 feet. The barrel is 3 feet high. Calculate (*a*) the volume of the barrel in cubic feet and (*b*) the number of gallons of liquid the barrel can hold.

Solution:

Step 1. Calculate the volume in cubic feet.

$$\text{Volume, in cubic feet} = \pi \times R \times R \times H$$

$$\text{Volume, in cubic feet} = 3.1416 \times 1.25 \times 1.25 \times 3$$

$$\text{Volume, in cubic feet} = 14.726$$

Step 2. Calculate the gallons of liquid the barrel can hold. (The cubic measure table shows that a cubic foot equals 7.4805 liquid gallons.)

$$\text{Barrel's liquid gallon capacity} = \text{number of cubic feet in barrel} \times \text{gallons (liq.) per cubic foot}$$

$$\text{Barrel's liquid gallon capacity} = 14.726 \times 7.4805$$

$$\text{Barrel's liquid gallon capacity} = 110.16$$

Calculations of cubic measure are frequently used to determine water and natural gas utility use and to do construction calculations involving cubic yards of gravel and concrete.

METRIC MEASUREMENT CONVERSIONS

The three principal base units of metric measurement are the *meter* (m) for length, the *liter* (L) for capacity, and the *gram* (g) for weight or mass.

Conversion from one unit to another within the metric system is far easier than within the English system. For instance, to convert English system units of length, inches are multiplied by 12 to convert to feet, feet are multiplied by 3 to convert to yards, and so on. In the metric system, the basic unit of length, the meter, is multiplied by 10 or a power of 10 to get a larger unit of measurement and divided by 10 or a power of 10 to get a smaller unit of measurement. This decimal system, using a base of 10, is used consistently for all units of metric measurement.

In the metric system, prefixes are added to the base unit to identify an increase or decrease from the base by a power of 10. For instance, 10 meters is called a dekameter, 100 meters is called a hectometer, and 1,000 meters is called a kilometer. One-tenth of a meter is called a decimeter, one-hundredth of a meter is called a centimeter, and one-thousandth of a meter is called a millimeter.

Metric prefixes, their symbols, and their numerical values in relation to the base unit are shown in Table 13-8. The procedure for converting from those units of measurement to the base unit

TABLE 13-8 METRIC SYSTEM PREFIXES, SYMBOLS, NUMERICAL VALUES, AND CONVERSION PROCEDURES

Prefix	Abbreviation	Numerical value in relation to base unit*	to convert these to the base unit*	Follow this procedure
milli	m	0.001	millimeter (mm) milliliter (mL) milligram (mg)	Divide by 1,000
centi	c	0.01	centimeter (cm) centiliter (cL) centigram (cg)	Divide by 100
deci	d	0.1	decimeter (dm) deciliter (dL) decigram (dg)	Divide by 10
deka	da	10	dekameter (dam) dekaliter (daL) dekagram (dag)	Mulitiply by 100
hecto	h	100	hectometer (hm) hectoliter (hL) hectogram (hg)	Multiply by 100
kilo	k	1,000	kilometer (km) kiloliter (kL) kilogram (kg)	Multiply by 1,000

*The base units are the meter, liter and gram. To convert *from* the base unit of meter, liter, or gram, follow the procedure opposite from that shown.

also is shown. To convert from one metric unit of measurement to the next higher one, divide by 10 (as when converting from milligrams to centigrams). To convert to the next lower unit, multiply by 10 (as when converting from hectoliters to kiloliters).

EXAMPLE Convert the following metric measurements to other metric values.

	Measurement	Convert to
(*a*)	2 kilometers	meters
(*b*)	3,000 meters	kilometers
(*c*)	600 millimeters	meters
(*d*)	70 centimeters	millimeters
(*e*)	1 gram	centigrams
(*f*)	1 gram	kilograms
(*g*)	40 hectoliters	liters

Solution: Use Table 13-8 as a guide.

	Measurement	Converted to	Procedure
(*a*)	2 kilometers	2,000 meters	2 × 1,000
(*b*)	3,000 meters	3 kilometers	3,000 ÷ 1,000
(*c*)	600 millimeters	.6 meter	600 ÷ 1,000
(*d*)	70 centimeters	700 millimeters	70 × 10
(*e*)	1 gram	100 centigrams	1 × 100
(*f*)	1 gram	.001 kilogram	1 ÷ 1,000
(*g*)	40 hectoliters	4,000 liters	40 × 100

CONVERSION BETWEEN ENGLISH AND METRIC SYSTEMS

Conversion between English and metric systems of measurement can be performed by using the accompanying conversion tables.

Measurement of Length

The basic metric measurement of length is the meter, which is slightly longer than a yard. Other measurements of length are given in Table 13-9.

TABLE 13-9 ENGLISH-METRIC CONVERSION TABLE, LINEAR MEASURE

Metric to English			English to Metric		
1 millimeter	=	.03937 inch	1 inch	=	2.540 centimeters
1 centimeter	=	.3937 inch	1 foot	=	0.304j8
1 decimeter	=	.3281 foot	1 yard	=	.9144 meter
1 meter	=	39.3701 inches	1 rod	=	5.0292 meters
1 meter	=	3.2808 feet	1 mile	=	1.6093 kilometers
1 meter	=	1.0936 yards			
1 dekameter	=	1.9884 rods			
1 kilometer	=	.6214 mile			

EXAMPLE Make the following conversions between the English and metric systems.

	Convert from	Convert to
(*a*)	240 miles	kilometers
(*b*)	308 kilometers	miles
(*c*)	100 yards	meters
(*d*)	60 centimeters	inches
(*e*)	50 decimeters	feet

Solution: Use Table 13-9 as a guide.

	Converted from	Converted to	Procedure
(*a*)	240 miles	386.23 kilometers	240 × 1.6093
(*b*)	308 kilometers	191.39 miles	308 × .6214
(*c*)	100 yards	91.44 meters	100 × .9144
(*d*)	60 centimeters	23.62 inches	60 × .3937
(*e*)	50 decimeters	16.41 feet	50 × .3281

Measurement of Capacity

The liter, which is similar in size to a quart, is used to measure both liquid and dry capacity. Each is identified in Table 13-10.

TABLE 13-10 ENGLISH-METRIC CONVERSION TABLE, CAPACITY MEASURE

Liquid measure	
Metric to English	English to metric
1 liter = 2.1134 pints	1 pint = .4732 liter
1 liter = 1.0567 quarts	1 quart = .9463 liter
1 liter = .2642 gallon	1 gallon = 3.7853 liters
Dry measure	
Metric to English	English to metric
1 liter = 1.8162 pints	1 pint = .5506 liter
1 liter = .9081 quart	1 quart = 1.1012 liters
1 liter = .2270 gallon	1 gallon = 4.4048 liters

EXAMPLE Make the following conversions between the English and metric systems.

	Convert from	Convert to
(*a*)	3 pints (liquid)	liters
(*b*)	12 liters (liquid)	gallons
(*c*)	14 quarts (dry)	liters
(*d*)	8 gallons (liquid)	liters
(*e*)	6 liters (dry)	pints

Solution: Use Table 13-10 as a guide.

	Converted from	Converted to	Procedure
(*a*)	3 pints (liquid)	1.42 liters	3 × .4732
(*b*)	12 liters (liquid)	3.17 gallons	12 × .2642
(*c*)	14 quarts (dry)	15.42 liters	14 × 1.1012
(*d*)	8 gallons (liquid)	30.28 liters	8 × 3.7853
(*e*)	6 liters (dry)	10.90 pints	6 × 1.8162

Measurement of Weight

The gram is the basic metric unit of weight. However, since a gram is only 3.5 percent of an ounce, it is too small for ordinary use and the kilogram (1,000 grams) is commonly used instead. As points of reference, a standard-size paper clip weighs about $\frac{1}{2}$ gram and a dime weighs slightly more than $2\frac{1}{4}$ grams. *Troy weight* (12 ounces to a pound) is used to measure precious metals and gems. *Avoirdupois weight* (16 ounces to a pound) is used for most other goods. See Table 13-11.

TABLE 13-11 ENGLISH-METRIC CONVERSION TABLE, WEIGHT (OR MASS) MEASURE

Avoirdupois weight					
Metric to English			English to metric		
1 gram	=	.03527 ounce	1 ounce	=	28.3495 grams
1 kilogram	=	2.2046 pounds	1 pound	=	453.5924 grams
1 metric ton	=	1.1023 short tons	1 pound	=	.45359 kilogram
1 metric ton	=	2,204.62 pounds	1 short ton (2,000 pounds)	=	.9072 metric ton
			1 short ton	=	907.18 kilograms
Troy weight					
Metric to English			English to metric		
1 gram	=	.03215 ounce	1 ounce	=	31.104 grams
1 kilogram	=	2.6792 pounds	1 pound	=	373.242 grams
			1 pound	=	.37325 kilogram

EXAMPLE Make the following conversions between the English and metric systems.

	Convert from	Convert to
(*a*)	8 pounds (avoirdupois)	kilograms
(*b*)	680 grams (troy)	ounces
(*c*)	4,650 pounds (avoirdupois)	metric tons

	Convert from	Convert to
(*d*)	7 kilograms (avoirdupois)	pounds
(*e*)	14 ounces (avoirdupois)	grams

Solution: Use Table 13-11 as a guide.

	Converted from	Converted to	Procedure
(*a*)	8 pounds (avoir.)	3.629 kilograms	8 × .45359
(*b*)	680 grams (troy)	21.86 ounces	680 × .03215
(*c*)	4,650 pounds (avoir.)	2.11 metric tons	4,650 ÷ 2,204.62
(*d*)	7 kilograms (avoir.)	15.43 pounds	7 × 2.2046
(*e*)	14 ounces (avoir.)	396.89 grams	14 × 28.3495

Measurement of Temperature

The metric thermometer, called the *Celsius thermometer,* uses a scale on which zero degrees is the freezing point and 100 degrees is the boiling point. The English system measurement of temperature is the *Fahrenheit thermometer,* on which 32 degrees is the freezing point and 212 degrees is the boiling point. Normal body temperature is 98.6 degrees Fahrenheit (F) or 37 degrees Celsius (C).

To convert from one of the temperature scales to another, use these formulas:

To convert from Fahrenheit to Celsius:

$$\text{Celsius} = \tfrac{5}{9}(F - 32)$$

To convert from Celsius to Fahrenheit:

$$\text{Fahrenheit} = (\tfrac{9}{5}C) + 32$$

EXAMPLE 1 A bank's time and temperature sign shows the temperature to be 30 degrees Celsius. How many degrees Fahrenheit is that?

Solution:

$$\text{Fahrenheit} = \left(\frac{9}{5}C\right) + 32$$

$$\text{Fahrenheit} = \left(\frac{9}{5} \times 30\right) + 32$$

$$\text{Fahrenheit} = \left(\frac{9}{5} \times \frac{30}{1}\right) + 32$$

$$\text{Fahrenheit} = \frac{270}{5} + 32$$

$$\text{Fahrenheit} = 54 + 32$$

$$\text{Fahrenheit} = 86 \text{ degrees}$$

EXAMPLE 2 Yesterday's high temperature was 80 degrees Fahrenheit. How many degrees Celsius is that?

Solution:

$$\text{Celsius} = \frac{5}{9} \times (F - 32)$$

$$\text{Celsius} = \frac{5}{9} \times (80 - 32)$$

$$\text{Celsius} = \frac{5}{9} \times 48$$

$$\text{Celsius} = \frac{5}{9} \times \frac{48}{1}$$

$$\text{Celsius} = \frac{240}{9}$$

$$\text{Celsius} = 26.67 \text{ degrees}$$

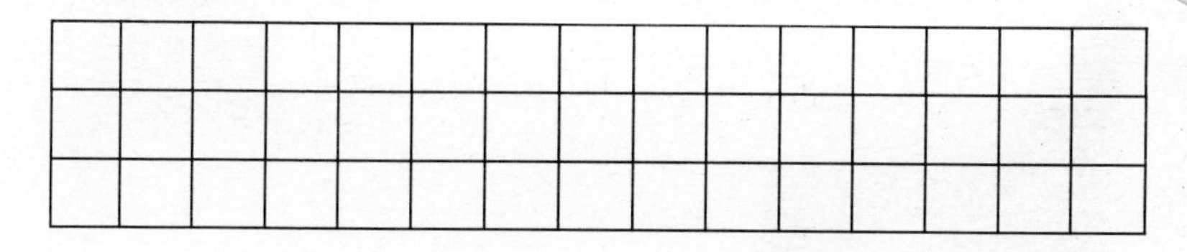

It's Fun to Experiment

The information in *Everyday Math Made Easy* provides you with the tools to make a large number of calculations that you will encounter in your day-to-day activities. Now that you have the tools, put them to use. Experiment.

Before obtaining that home loan, determine the difference in monthly payments between a 15-, 20-, and 25-year loan and calculate the difference in total amount required to repay each loan. Figure out what that new appliance you'd like to buy will cost to operate. Calculate what that new carpet or wallpaper will cost—maybe you *can* afford it after all. Prepare an income statement, balance sheet, and personal budget. The results might be interesting!

You can perform all of these and hundreds of other calculations easily and quickly by using *Everyday Math Made Easy* as a guide. Keep it handy and refer to it often.